POND FISH ECOLOGY

AND

ECONOMICS

POND FISH ECOLOGY AND ECONOMICS

D. R. Khanna
&
S. Singh

2013
Daya Publishing House
A Division of
Astral International Pvt. Ltd.
New Delhi - 110 002

First Printed: 2002
Reprint: 2013

ISBN 978-81-7035-281-5

Published by : **Daya Publishing House**
A Division of
Astral International Pvt. Ltd.
81, Darya Ganj, Near Hindi Park,
Delhi Medical Association Road,
New Delhi - 110 002
Phone: 011-4354 9797,23278134
Fax: +91-11-2324 3060
E-mail: info@astralint.com
Website: www.astralint.com

Laser Typesetting : **Classic Computer Services**
Delhi - 110 035

Printed at : **Salasar Imaging Systems**
Delhi - 110 035

PRINTED IN INDIA

Dedicated

to

Nature

CONTENTS

NOTATIONS AND ABBREVIATIONS

Symbol	Meaning
$^{\circ}$	Degree
'	Minute
m^2	Square Meter
m^3	Cubic Meter
$^{\circ}C$	Degree Centigrade
$\pm$	Standard-deviation
@	On Coast Of

Abbreviation	Expansion
Lat	Latitude
Long	Longitude
T.S.	Total Solids
T.S.S.	Total Suspended Solids
T.D.S.	Total Dissolved Solids
J.T.U.	Jackson's Turbidity Unit
D.O.	Dissolved Oxygen
BOD	Biochemical Oxygen Demand
COD	Chemical Oxygen Demand
ppm.	Parts Per Million
MPN	Most Probable Number
SPC	Standard Plate Count
CPUE	Catch Per Unit Effort
Ha	Hectare

CHAPTER - 1

INTRODUCTION

"Water is the best of all things"

by *Pindar*

If the question is asked, whether life could exist on a certain planet, the first question a biologist asks, Is water in liquid form available there?

If it is not, life as we know it could not exist, all the reactions of life are carried out in aqueous solution.

If any single compound should be selected as characteristics of life, it is water.

Nearly more than 75 percent of the earth's surface covered by water either as ocean, or fresh water therefore, earth is some times called a "water planet" water is in sea, rivers ocean, lakes, pond, streams etc.

It exists in all three form viz. solid, liquid, and gas, most of the earth's surface water is in sea, the oceans, which contains about 35 parts per thousand of dissolved salts of the remained, most of fresh with a salt content of 0.2% is found either in lakes and ponds, or in rivers and streams. Fresh water is also available in form of rain, snow, dew etc.

Evaporation of water from ocean, cloud formation, and precipitation are responsible for worldwide water circulation through hydrological cycle. Life was first organized in water, water is one of the primary necessity of living and nonliving cycles occurring in the universe.

Water is used in every part of day to day life. In natural conserving surface water is a habitat of wild life and to the general people recreational aspects are important apart from drinking water.

Chemically water is a monoxide of hydrogen (HOH). If feebly dissociates into hydrogen and hydroxyl ions, the ionization product $(H^+) \times (OH^-)$ is about 10^{-14}, water is a colorless liquid and possesses a high dielectric constant (81) and therefore, salts get highly ionised when dissolved in water but not so in other solvents, due to this property, water is known as universal solvent present on the earth, that is why the natural water contains many salts in dissolved form.

Water has high specific heat and this property of high heat capacity of water is functionally important to aquatic organisms.

Water possess a highest heat of fusion and heat of evaporation, collectively known as "latent heat", of all known substance that are liquid at ordinary temperature. The latent heat of water moderates the temperature of the biosphere, it also plays an important role in evaporation of water and its condensation as rain and as dew in the hydrological cycle.

Water is a poor thermal conductor as compared to metals. It has high viscosity which allows organisms to swim using relatively simple moments. It also protects the aquatic animals and plants from the mechanical disturbances.

The surface tension of water is greatest among all common liquids except mercury, the very high surface tension is useful in the movement of water through and into the organisms.

Water is also a boyont medium, organisms can exist with in it without specialized supportive structures.

Natural waters also contains ions such as Na^+, K^+, Mg^{+2}, Cl^-, SO_4^{-2}, PO_4^{-3}, NO^{3-}, CO^{3-} and HCO_3^- these all ions are responsible for the salinity of the water.

Most of gases, especially those that are essential for life are soluble in water, oxygen dissolved in water is used by most living organisms for respiration, Carbon-di-oxide produced by decomposition of organic matter and the respiratory activity of aquatic plants and animals, CO_2 also combines with water to form carbonic acid (H_2CO_3) which influences the pH or H^+ ion concentration of water.

On the basis of dissolved salts in water, water may be classified as fresh water and marine water (salty water).

The study of fresh water ecosystems of all kinds, lakes, reservoirs, streams, ponds, marshes bogs etc. is known as limnology.

Ponds are small bodies of water in which the littoral zone is relatively large and the limnetic and profundal regions are small or absent. Stratification is of major importance.

Ponds may be found in most region of adequate rainfall. They are continuously being formed for example: as a stream shifts position leaving formed bed isolated as a body of standing water of "oxbow". Such ponds can be designated as flood plain ponds and are subject to flooding during high water because of accumulation of organic materials on flood plain, small flood plain ponds may be quiet productive in the pie dent region of the south-east, where glacial lakes and ponds are lacking, we have found that flood plain ponds are very favourable sites for class study since they contain a rich and varied life and illustrate various principles of pond ecosystem. Natural ponds are also formed in lime-stone regions, when depressions or "sinks" develop due to solution of under lying strata.

Temporary ponds, which are dry for part of the year, are especially interesting and support a unique community organisms, in such ponds must be able to survive in dormant stage during dry period, or be able to move in and out of ponds, as can amphibious and adult aquatic insects. The fairy shrimps *(Eubranchiopoda)* or especially remarkable crustaceans which are well adopted to temporary ponds. The eggs survive in dry soil. For many months, whereas development and reproduction occur in short time late winter and springs while water is present.

Even though a temporary pond contains water only for few weeks, a seasonal succession of organisms may occur, thus enabling a surprisingly large varieties of organism to utilize a very limited amount of physical habitat.

Artificial ponds of course are the result of domming of streams as basis by man, or by animal, such as the beaner prior to about 1920, most man-made ponds in the United State were "mill ponds" formed by domming sizable streams for the purpose of providing power for small mills.

At the present time a very large number of "form ponds" are being constructed which differ from "mill ponds" in that stream water is largely detoured around the pond, in order to prevents loss of nutrients and erosions, or the pond is constructed in basin without a permanent stream. Such pond have relatively little water flowing through them, and are after artificially fertilized and managed so as discourage rooted vegetation. Thus producers are all phytoplankton and in a sense, food chain relations resemble those of lakes.

However such ponds are rarely deep enough for stratification. A form pond can be much more efficient than an unmanaged pond as mill pond from the stand point of production of a large bio-mass of fish, but it is less interesting to the naturalist because the variety of organisms is reduced in favour of a large numbers of desired species.

Bearer ponds are a characteristic feature of wild less areas of the north and of mountain of west. A bearer pond generally has a short ecological life history, as such ponds are abandoned when the supply of food trees in the vicinity becomes reduced.

Thus, under primeval conditions, the bearer was very important factor in opening up the forests and maintain both terrestial and aquatic several stages.

In the South-eastern states and Texas, bearer do not always construct ponds but after live in holes in the band of streams and thus become essentially stream animals.

Let us consider ponds as a whole as an ecosystem. The inseparability of living organisms and the non-living environment is at one apparent with the first sample collected. Not only is the pond a place where animals and plants live, but the animal and plants make the pond what it is, thus a bottle full of pond water organic compounds, some of the larger animals and plants can be separated from the sample for study of counting but it would be difficult to completely separate the myriad of small living things from the non-living matrix without changing the character of the fluid. Thus, one could autoclave the sample of the water or bottom mud so that only non-living material remained, but this residue would then no longer be pond water or pond soil but would have entirely different appearances and characteristics.

Temperature influences the number and types of saprophytes, reaeration, O_2 stability and biochemical reaction taking place in a stabilization pond (Davis 1932).

To estimate the total bioactivity of lake, reservoir of pond, it is necessary to determine the magnitude of primary production and efficiency of energy utilization at different trophic levels.

For the purposes, the estimation of O_2 content in all euphoric zone of any aquatic habitat and indirect estimate of the algal matter seem to after rapid procedure for determining, through approximately rate of photosynthesis and energy flow of different trophic levels at various in a year.

Our country is bestowed with bountiful natural resources in the form of extensive coast line, river systems, estuaries, ponds, tanks, lakes, reservoirs and bells, exposed to the warm, tropical climate, these water bodies are extremely productive and they harbour enviable spectrum of fish genetic resources. Optimum utilization of these resources and lead to a many fold increase in inland fish production earning the country a place among the top in land fish producing nations of the world.

Inland fish production in the country witnessed a six fold increases from 0.22 million ton in 1950-51 to 1.30 million ton in 1987-88, the overall growth rate in India fish production is almost on a par with a global four fold increase. A quantum jump in inland fish production occurred during the 6th and 7th plan period which witnessed overall growth rate more than 5 per cent per annum. Against a declining trend noticed in the world as a whole. The projectable achievable target of fish production by 2000 A.D. is 8.0 million ton, of which the share of inland fisheries is estimated at 4.5 million ton. The country as 7.3 lakhs hectare of ponds and tanks that can be used for fresh water aquaculture, though only 1.50 lakhs hectare are being utilized at present. The remaining 6 lakh hectare either remain fallow or produce fish at subsistence level.

Therefore scientific utilization of water resources, the study of phytoplankton and physico-chemical characteristics of water are of primary interest.

In ecological study of Padali Gujjar Pond during pre-autumn, autumn and post-autumn seasons during 1998-99.

The "Padali Gujjar Pond" is situated at village Padali, Roorkee in district Haridwar of Uttaranchal.

CHAPTER - 2

PHYSICO-GEOGRAPHY

Roorkee, a moderate size town in the district of Hardwar (Uttaranchal), is located on both of the banks of upper Ganga Canal, which takes off at Hardwar, 30 Km away. About 30 Km South of Shivalik and about 180 Km North of Delhi, it is linked by rail to most important towns such as Delhi, Bombay, and Calcutta and is well connected by roads. The nearest airport is about 65 Km away at Dehradun and a halipad in ground of Bengal Engineers, apart from an airstrip facility at B.H.E.L., Hardwar

The town is 268 meter above mean sea level (longitude 77 degree 53 min East and latitude 29 degree 51 min North) and has a cold winter. The summer months, though hot, are moderated by the effects of Shivaliks. The rain fall mainly takes between July and September, the average being 1050 mm (Table 2.1). Becides the University, which has 150 hectare campus, the town is an important center for engineering, the Central Building Research Institute and the, U.P. Irrigation Research Institute with the activity of Bengal Engineers, National Institute of Hydrology, U.P. Irrigation Design Organisation being located here.

The town has a total population of approx. 90,000 (according to census 1991) and is surrounded by Hardwar, Saharanpur and Muzzafarnagar Districts.

Pond is situated at village Padali-Gujjar, which is 3.5 Km. away from bus stand and 0.5 Km. away from railway station on Paniyala road (Fig. 2.1).

At the western bank of the pond, Padali-Gujjar village is present and the Eastern bank of the pond touches to upper Ganga Canal. Between canal and pond a road is present which goes from Roorkee to Manglore along the canal, World bank colony also situated at the other bank of canal and the southern bank touches the fields of the farmers, various types of crops are grown in these fields.

The plot no of the pond is 633 and area is 72 bhiga (4.8 ha) (according to Government record). The source of water in pond is rain water and ground water from canal. Padali-Gujjar pond is at height of 268 meter from the mean sea level (longitude 77 degree 53 minute East and latitude 29 degree 51 minute North). The length of pond is 650 meter and width is 450 meter, the maximum depth of pond is 4.8 meter and minimum 0.6 meter, mean depth is 2.7 meter. The surface area of Padali-Gujjar Pond is 2,92,500 square meter, catchment area is 2,92,400 square meter and volume of water is 7,89,750 cubic meter. (Fig. 2.2 and Table 2.1)

Pond receives little domestic sewage from drains of adjacent village, and in rainy season fertilizers, pesticides are also swept with rain water and mixes in pond.

The study was conducted during three different seasons viz Pre-autumn, Autumn and Post-autumn seasons of year 1998-99 at the three sampling sectors (Figs. 2.3 to 2.5).

Sampling Sector-I

Sampling sector-I is situated at the western bank of the pond near the village. The depth of water is 0.45 meter to 0.61 meter at mid, this sector receives little sewage from village drains, this sector also used by villagers for washing their tractors and bullock carts.

The bed of pond in this sector is clayey.

Sampling Sector-II

Sampling sector-II is situated at the mid-point of the pond, the depth of water in this sector is about 1.76 meter, at the mid of this sector a elevation of 40 × 60 meter also presents, watchman and fishermen uses a wooden boat from bank to elevation. A hut is also present for resting purpose. The bed of pond in this sector is also sandy and clayey.

Sampling Sector-III

Sampling sector-III is situated at the eastern bank of the pond, which is adjacent to upper Ganga Canal. The depth of water in this sector ranges from 2.42 meter to 4.84 meter. In this sector some washermen also washes clothes.

The bed of pond in this sector is clayey.

Table 2.1 : Physico-Geographical Features of Padali-Gujjar Pond Roorkee

S. No.	*Parameters*	*Measurements*
1	Height (from sea level)	268 meter
2	Latitude	29 degree 51 minute North
3	Longitude	77 degree 53 minute East
4	Length of pond	650 meter
5	Width of pond	450 meter
6	Minimum depth	0.6 meter
7	Maximum depth	4.8 meter
8	Mean depth	2.7 meter
9	Surface area	292500 sq. meter
10	Catchment area	292400 sq. meter
11	Water volume	789750 M^3
12	Minimum air temperature	9°C
13	Maximum air temperature	37°C
14	Mean air temperature	23°C

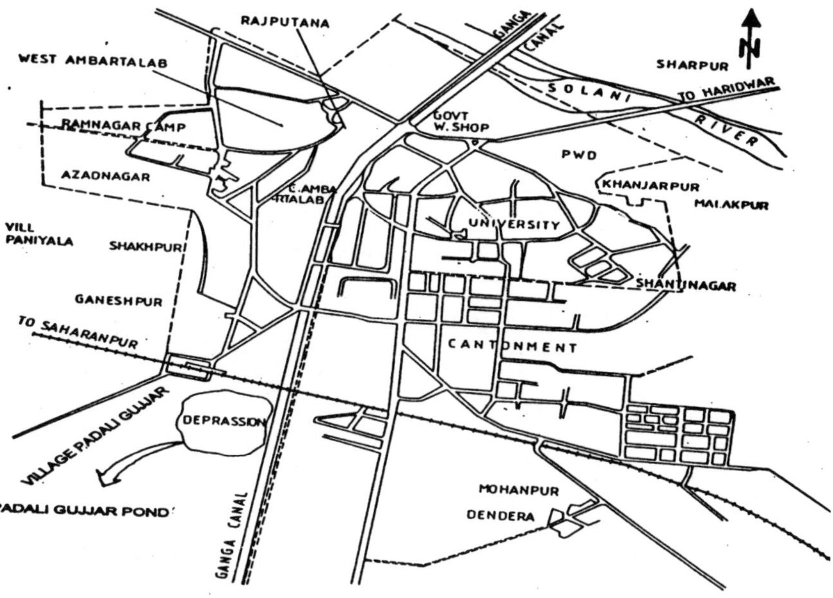

Fig. 2.1 : Plan of Roorkee Town : Indicating Study Site

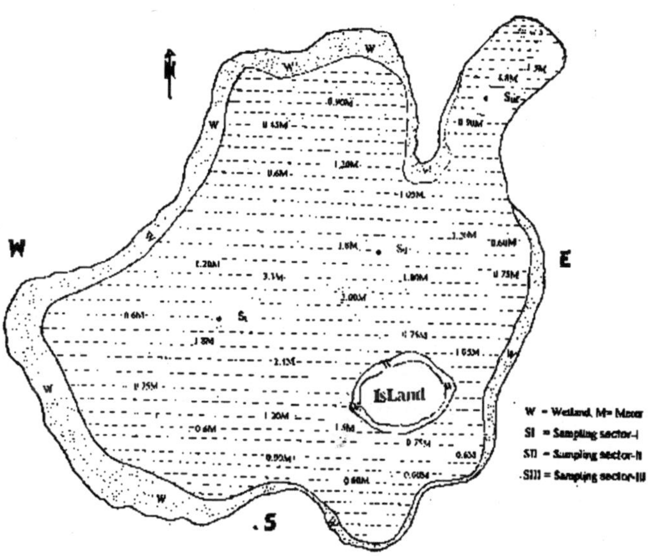

Fig. 2.2 : Map of Padali Gujjar Pond at Roorkee

Fig. 2.3 : Sampling Sector - I

Fig. 2.4 : Sampling Sector - II

Fig. 2.5 : Sampling Sector - III

CHAPTER - 3

REVIEW OF LITERATURE

History reveals that the beginning of civilization always took place near the water sources.

Hippocratus (460 BC to 354 BC) the father of medicine, stated that "water contributes much of health" and asserted that the rain water should be boiled and filtered before its use, otherwise it would have a bad smell and cause harshness. The evidence that the people of Egypt were probably the first in the world to culture fish, a far back as 2500 BC comes from a pictorial engraving on an ancient egyptician tomb showing, tilopia being fished out of an artificial pond (Marr *et al.* 966).

In the beginning the fish practice of fish culture in ponds, probably had its sole purpose of keeping fish caught in lakes and rivers alive for some time. Carp culture was widely spread in China in 2000 BC.

Fish culture in Indian continent is undoubtedly 100 years old. The importance of fish culture as a source of food production was then driven home more realistically and emphasis was layed on the need for extending fish cultural activities to all the parts of the country with a view to developing the industry on scientific lines both in the private and public sectors.

F.A. Forel (1892) of Switzerland may be considered as the founder of the modern limnology, whose original works on Swiss lakes was elaborated in three volumes the classic entitled. "Le Lemane" (lake of Geneva). He was the first to study and elaborate the physical and chemical and biological characteristics of lakes. The first text on limnology was also by Forel as "Hand buch der Senakunde Allegemeine limnologie".

The most detailed researches on physico-chemical and biological characteristics were carried out by Birge and Judey (1911). Who reports on Lake Wiscosin.

The most recent treatise on all aspects of limnology in the 3-volume, "Treatise on Limnology" by Hutchinson (1957,1967,1976), other work in area by Welch (1948), Ward and Whipple (1966) (Fresh water Biology revised by Edmonson), Reid (1961). Investigated ecology of Inland waters and estuaries at New York. Peenak (1962) investigated quantitative zooplankton sampling in littoral vegetation areas. Hynes (1970) observed the biology of polluted water. Needham (1972) given a guide to study of fresh water biology.

Thus in last six decades, striking advances have been made in the field of limnology in Europe and America.

Analysis of water samples for physical, chemical and biological parameters forms the base of limnological investigations. The main methods given by Welch (1952), Ruttner (1962), Sladaeck (1963), Golterman (1969), Hynes (1970), Hart (1974), Lind (1974), Prescott (1972), Macan (1974), Cook (1977).

In India, the first work on hydrology and hydrobiology was of a pond in Indian museum compound at Calcutta by Pruthi (1933), his studies on the bionomics of fresh waters of India. Faruqi (1943) reported method of catching fish in united provinces and the scope of establishment of Inland fishes in these provinces.

In Pakistan Ahmad (1952) has reported various stocking combinations of indigenous major carps based on the ecological conditions of the pond water.

Nakamura (1954) has reported that in the farms ponds of Japan, the basic species combinations in practice is common carp, crucian carp, silver carp and big head, recently been introduced in Japan.

Chacko and Krishnamurthy (1954) studied on the plankton of three fresh water ponds in Madras city.

Cultured fishes are often classified according to the trophic niche, occupy in a water body. The fishes have been grouped into (1) plankton eating or surface feeder such as *Catla-catla, Hypophthalmichthys-moltrix.* (2) Column or mid feeder fishes such as *Labeo-rohita.* (3) Bottom feeder such as *Cirrhinus-mirgala, C. reba, Labeo calbuso.* (4) Aquatic macrophytic such as Grass carp *(Ctenopharyngodon-idella).* Fish belonging to surface, column and bottom feeding categories have been sub-grouped according to character of the food they consume, into herbivores, carnivores and omnivores (Das and Moitra 1955).

The term weed fishes is used here to include all species of uneconomic, small sized fishes that naturally occurs or accidently introduced in the pond along with carp spawn. The role of predatory and weed fish in the ponds has been discussed by Alikunhi *et al.* (1956 a), Ibrahim (1957), Choudhary (1960). Predatory fishes are harmful to cultured major carps. Not only they compete with the later for food and space but also directly prey on them. Mahajan (1961) investigated the fish fauna of Muzaffarnagar district. Byars (1960) made a study on a fresh water pond in Newzealand.

Some more important consideration for attaining high fish production in stocking ponds are species combinations, there ratio, and rate of stocking. Every pond irrespective in of whole manner and can be managed can support a fish biomass only up to a certain carrying capacity (Yashouv, 1959) or the maximum standing crop (Hickling, 1962-71). The carrying capacity of the pond is intimately related to their nature, productivity, and management practice adopted, including fertilization, artificial feeding, the species stocked and their stocking rate. Zafar (1964) make studies on the ecology of algae in certain fish ponds in Hyderabad.

Pant (1970) has reported fish fauna of Kumaon hills. Munnawar (1970) make limnological studies on fresh water ponds of Hyderabad. Bhuyan (1970) took qualitative analysis of some ancient water tanks in Sibsagar (Assam). Verma and Shukla (1970) studied the physico-chemical parameters of Kamla Nehru tank Muzaffarnagar. Das and Akhtar (1970) reported the micro-organism of Dal lake, Kashmir. Sahagal (1973) made a study on fish and fisheries of Aligarh with notes on the fish catches. Singh (1974) reported fishes of Moradabad (U.P.). Natarajan (1974) investigated the ecology and fish culture in reservoir of India with special reference to Rihand sagar, Juna sagar, Bhavani sagar, Konar and Tilaiya.

Dabron (1976) observed physical and chemical factors of a vernal temporary pond in western Canada. Yousuf and Quadri (1978) studied the seasonal variation in the physico-chemical factors of sub-topical lakes of Kashmir.

Palmer (1980) has also noted species of *Anacystis, Chlamydomonas, Chlorella, Euglena, Melosira, Navicula, Nitzschia, Oscillatoria, Pandorina, Stiqeadonium* and *Synedra* etc. as the pollution indicator algae.

Yang (1982) observed fish culture and reservoir management in the republic of Singapore.

Excepting a few studies example Balone and Coche (1974) for lake Kariba in central Africa and Scheimer (1983) for Parakrama samudra in Srilanka, both executed by FAO.

Bagde and Verma (1985) reported physico-chemical characteristics of a shallow tropical JNU Lake at New Delhi and their seasonal patterns and periodicities were studied from August (1979) to July (1988). Singh and Sahai (1986) studied seasonal variations in phytoplankton in "Jal wania Pond" in Gorakhpur and observed maximum phytoplankton population during summer (113700/1). The phytoplankton population during summer included Cyanophyceae (36.6%), Chlorophyceae (13.4%), and Bacillariophyceae (50.10%). Sahai *et al.* (1986) studied the water quality of Chilka lake at Gorakhpur, which receives effluent of fertilizer factory, the author found that in lake water pH ranged from 9.8 to 11.6 (alkalinity) and 2.9 to 3.2 (acidity), dissolved oxygen 15.0-18.0 ppm and BOD 120-150 ppm. Dutta *et al.* (1987) recorded the seasonal abundance of Rotifers in perennial fresh water pond at Calcutta. Singh and Mahajan (1987) study the phytoplankton and water chemistry of Rewalsar pond and Renuka lakes, Himachal Pradesh.

Lake Rangasagar is a two century old highly polluted lake of Udaipur (Rajasthan). The lake receives enormous quantity of organic matter, which has accelerated the process of eutrophication The progressive eutrophy of the lake is revealed by planktonic community succession and limnological characteristics through the basic structure is suggestive of oligotrophy Rao and Durve (1989). Tripathi (1990) made investigation on fresh water aquaculture in India. Prasad (1990) reported semiemperical method to determine the photo-synthetic activity and energy flow in upper Lake, Bhopal.

Adholia (1991) study the gross production, net production and respiration rates of Mansarovar reservoir, Bhopal. A pisciculture water sheet, were studied from February to December (1990) by light and dark bottle method.

Chopra and Patrik (1991) studied on water pollution at Rishikesh, Joshi *et al* (1991) studied on hydrobiology of river Bhagirathi at Tehri Gharwal. Talwar and Jhingran (1991) reported Inland Fishes of India and adjacent countries.

Tombi Singh (1992) studied an eco-biology of stream fishes in special reference to culturable possibilities.

Khanna (1993) reported the pollution aspects of river Ganga at Haridwar. Gaur (1992) has been concluded that raw water supply from Keetham lake is fit for industrial as well as potable purposes, but it cannot serve as scarcity alternate to Agra water supply. Kumar (1992) recorded the effect of methyl parathion on primary production of phytoplankton of pond ecosystem on seasonal ecological micro-organism from sub-tropical pond of Jammu.

Sharma *et al.* (1993) study the abiotic profile of Garhwal Himalayan lake, the Deoria Tal.

Verma and Mohanty (1995) study the phytoplankton and its correlation with physico-chemical parameters of Danmukundpur pond. It was observed that during summer season Cyanophyceae and during winter Bacillariophyceae were maximum. pH showed a positive relationship with Bacillariophyceae and Cyanophyceae.

Dwivedi *et al.* (1995) made a limnological study in relation to fish production of two ponds at Patna.

Paka (1997) found the inverse relationship in solids and temperature.

Padmavati and Prasad (1997) study the impact of unwanted fish on the water quality, plankton and fish production in a carp culture in the envirous of lake Kallera. In Carp culture ponds one of the major factor responsible for poor production of Carp is the existence of unwanted fish which compete with cultured carp for food and basic environmental factors.

Khanna *et al.* (1997) investigated population of green algae in relation to physico-chemical factors of the river Ganga.

Jindal and Ghezta (1998) reported pollution status of water bodies of Chandigarh.

CHAPTER - 4

MATERIALS AND METHODS

Sampling Method

(A) Cleaning and sterilization of glass wares :-

Washing soda / chromic acid was used for cleaning the different glass wares.

(1) Sample bottles

300 ml capacity B.O.D. bottles made up of borosil were used . They were washed with washing soda and chromic acid and rinsed with tap-water followed by distilled water, the neck and stoper were wrapped by brown paper with the help of rubber bands. Sterilized the sample bottle in autoclave at 15 lbs pressure or 121°C for 15-20 minutes.

(2) Pipettes

Pipettes were washed and fitted with plug of cotton wool at the upper end wrapped in a paper and there sterilized in autoclave at 15 lbs pressure for 15-20 minutes.

(3) Petri-dishes

Petri-dishes were washed with washing soda as well as with chromic acid and rinsed with tap water followed by distilled water and than sterilized in oven at 160 to 180°C for 2-3 hours test tubes, conical flask, measuring cylinder, burette beaker: washed with washing soda as well as with chromic acid and rinsed with tap water followed by distilled water and then sterilized in oven at 160 to 180°C for 2-3 hours.

(B) Sample Collection

Sampling equipments:

Following equipments are used for sampling water samples.

(a) Sampling bottle.
(b) Jericans

Water samples was taken from 2 feet below the surface water by keeping and opening jericans at desired depth below the surface water. Caps of the canes are removed after dipping the cane at the desired depth and also closed in the water after filling up of cane.

(1) Sampling for dissolved oxygen

For dissolved oxygen determination, the samples were collected in (clean and sterilized 300 ml). BOD bottles and completely filled with the sample pressure by the addition of 1 ml of alkaline Kl and 1 ml of $MnSO_4$ solution at sampling spot.

(2) Sampling for other chemical parameters besides pH and CO_2

For chemical parameters the samples are collected in jericans of 2.5 liter capacity. The water samples are collected from a depth of 2 feet, samples are collected up to the top without leaving any space, so as to prevent the premature release of dissolved oxygen during the transit period.

(3) Sampling for biological analysis

The sample was collected by a plankton net attached with a test tube at the base. For this, a known amount of water passed through the plankton net and the water collected in the test tube. Transferred in a 50 ml plastic bottle, and preserved by adding 2-3 drops of formalene, and the bottle was labelled indicating date, time, and sampling sector no.

(C) Water Analysis Methods

The physico-chemical and biological analysis of water samples were made of all the collected sample following the methods out lined in:

"Standard Method for examination of Water and Waste water "APHA" (American public health association 15th ed. 1980) and biological method for water pollution studies by Trivedi and Goel (1984) and Mathur (a laboratory manual) and Saxena (1994)".

Samples were analysed for the following parameters:

(A) Physico-chemical parameters:

(1) Temperature
(2) Total solids (TS)
(3) Total suspended solids (TSS)
(4) Total dissolved solids (TDS)
(5) Turbidity
(6) Conductivity
(7) pH
(8) Dissolved oxygen (DO)
(9) Free CO_2
(10) Bio-chemical oxygen demand (BOD)
(11) Total alkalinity (T.A)
(12) Hardness
(13) Calcium (Ca)
(14) Magnesium (Mg)
(15) Sodium (Na)
(16) Potassium (K)
(17) Chlorides (Cl)
(18) Chemical oxygen demand (COD)

Heavy Metals

(19) Zinc (Zn)
(20) Iron (Fe)

(21) Manganese (Mn)
(22) Cobalt (Co)

(B) Biological Parameters

(1) Quantitative estimation of total plankton
(2) Qualitative estimation of various types of phytoplankton.
(3) Standard Plate Count (SPC)
(4) Most Probable Number (M.P.N.)
(5) Fishes

Methodology

(1) *Temperature*

The temperature of water measured by using a thermometer, Calibrated in °C scale. For measuring the temperature of open water body, the thermometer is dipped into the water as near about 6" and the reading is taken. Thermometer should not be taken out from the water body for reading temperature. To minimize the errors it is always essential to calibrate the thermometer with an another thermometer of known accuracy.

(2) *Total Solids (T.S)*

Total solids are determined as the residue left after the evaporation of the unfiltered sample.

Procedure

Take an evaporating dish (made up of silica, porcelain or platinum) of at least 100 ml capacity. Iginite at 550°C in a muffle furnace for about an hour, cool in a desiccator and weight. Evaporate 100 ml of unfiltered sample in the evaporating dish on hot plate having temperature not more than 98°C. Heat the residue at 103°C-105°C is a oven for one hour and take the final weight after cooling in a dessicator.

Calculations

$$\text{Total solids (ppm.)} = \frac{A\text{-}B \times 1000 \times 1000}{V}$$

Where,

A= final weight of the dish in gm.

B= initial weight of the dish in gm.

V= volume of sample taken in ml.

(3) Total Suspended Solids

Determine total suspended solids as the difference between the total solids and total dissolved solid.

Calculation

$$\text{T. S. S. (ppm.)} = \text{TS - TDS}$$

(4) *Total Dissolved Solids (TDS)*

Total dissolved solids are determined as the residue left after evaporation of the filtered sample.

Procedure

Evaporation dish is ignite and weight, and dried 100 ml of the filtered sample in a pre-weight evaporating dish at 103-105°C in hot air oven for one hour, and weight after cooling in a dessiccator.

Calculations

$$\text{Total dissolved solids (ppm.)} = \frac{\text{A-B} \times 1000 \times 1000}{\text{V}}$$

Where,

A = final weight of dish in gm.

B = initial weight of dish in gm

V = volume of sample in ml.

(5) *Turbidity*

The turbidity of water was measured with the help of **Jackson's candle turbidity meter.** It consists of three parts (a) candle, (b) holder, (c) calibrated glass tube. The glass and candle are supported in a vertical position, so that the center line of the tube may pass through central axis of the candle flame, the sample is poured into the glass tube slowly until the image of the candle flame disappear. The values are calculated as described by Mathur (1988).

(6) *Conductivity*

Electrical conductance is the ability of a substance to conduct the electric current. In water it is the property caused by presence of various ionic species.

It is generally measured with the help of a **conductivity meter** having a conductance cell containing electrode of platinum coated with Pt. black or carbon.

Procedure

It is measured with the help of conductivity meter. Conductance depends upon the area of the metallic electrodes and distance between them. The factor used to convert the observed conductance into conductivity is called cell constant. Take sample in a beaker and dip the electrode and set the dia at point and read out the readings.

Calculation

Conductivity (micro-mho) =Observed conductance × cell constant

(7) *pH*

pH is the negative logarithm of the hydrogen ion concentration in a solution pH of the water samples was measured with the help of pen digital pH meter.

Procedure

Sample taken into the beaker and after opening the cap, pH meter dipped into sample and on it. pH reading noted from digital screen of meter.

(8) *Dissolved Oxygen (D.O.) : (Wrinklers Iodometric Method)*

The manganous sulphate reacts with the alkali (KOH, NaOH) and make white precipitate of manganous hydroxide which in the presence of oxygen gets oxidized to brown colour compound. In the strong acid medium manganic ions are reduced by iodine ions which get converted to iodine equivalent to the original concentration of oxygen in the sample. The iodine can be titrated against sodium thio sulphate using starch as an indicator.

Regents

(a) Sodium Thio Sulphate Solution, 0.025N

Dissolve 6.205gm. of sodium thio-sulphate ($Na_2S_2O_3.5H_2O$) in previously boiled distilled water and make up the volume 1 litre. Add a pallet of sodium hydroxide as a preservative. Keep the solution in colobvgftr54red bottle.

(b) Manganous Sulphate Solution

Dissolve 100gm. of $MnSO_4.4H_2O$ in 200ml. of boiled distilled water and filter the solution.

(c) Alkaline Potassium Iodide Solution

Dissolve 100gm. of KOH (Potassium hydroxide) and 50gm. of KI (Potassium iodide) in 200ml. of previously boiled distilled water.

(d) Starch Indicator

1 gm. of starch in 100 ml. of warm (80-90°C) distilled water and add few drops of formaldehyde as preservative.

(e) Sulphuric Acid

Concentrated sp. gr. 1.84 :18M

Procedure

Fill the water sample in BOD bottle carefully, avoiding any kind of bubbling. No air should be trapped in bottle after the stopper is placed. Open the bottle and add in it 1 ml of manganous sulphate and 1 ml of potassium iodide solution by using separate pipettes. A precipitate will appear. Place the stopper and shake the bottle thoroughly. Keep the bottle for some time to settle down the precipitate. If the titration is to be prolonged for few days keep sample at this stage with the precipitate. Add 2 ml of concentrated sulphuric acid and shake the bottle thoroughly to dissolve the precipitate. Remove 100 ml of contents in conical flask for titration. Put a few drops of starch indicator and titrate against 0.025N sodium thio-sulphate and note the end point, When initial blue color turns to colorless.

Calculations

$$\text{Dissolved oxygen (ppm.)} = \frac{(\text{ml} \times \text{N}) \text{ of titrant} \times 8 \times 1000}{\text{ml of sample}}$$

(9) *Free CO_2*

Free CO_2 can be determined by titrating the sample using strong alkali (such as carbonate free NaOH) to pH 8.3. At this pH all the CO_2 converted into bicarbonates.

Reagents

(a) Sodium Hydroxide, 0.05N

Prepare 1 N NaOH solution, by dissolving 40 gm of NaOH in free CO_2 water (boiled) to make 1 lt. of solution diulte 50 ml of 1.0N NaOH to 1 litre distilled water standardized if with H_2SO_4, HCI or oxalic acid.

(b) Phenolpthalein indicator

Dissolve 1gm of phenolpthalein in 100 ml of 95% of ethanol and 100 ml of distilled water. Add .05N CO_2 free NaOH solution dropwise, until the solution turns faintly pink.

Procedure

Take 50ml of sample in a conical flask and add few drops of phenolpthalein indicator. If the color turns pink, free CO_2 is absent. If the sample remains colorless titrate it against 0.05N NaOH at the end point a pink colour appears

Calculations

$$\text{Free } CO_2 \text{ (ppm.)} = \frac{(\text{ml} \times \text{N}) \text{ of NaOH} \times 1000 \times 44}{\text{ml of sample}}$$

(10) Biochemical Oxygen Demand (BOD)

Biochemical oxygen demand is the measure of degradable organic matter present in a water sample and can be defined as the amount of oxygen required by the microorganism in stabilizing the biological degradable organic matter under aerobic conditions.

The principal of the method involves. measuring the difference of the oxygen concentration between the sample and after incubation for 5 days at 20°C.

Reagents : Same as DO

(11) *Total Alkalinity*

Total alkalinity is the measure of the capacity of the water to neutralize a strong acid. The alkalinity in the water is generally imparted by the salts of carbonates, bicarbonates, phosphates, nitrates, borates, silicates, etc., together with the hydroxyl ions in free state. However most of the water are rich in carbonates and bicarbonates with little concentration of other alkalinity imparting ions.

Total alkalinity, carbonates and bicarbonates can be estimated by titrating the sample with strong acid (HCl or H_2SO_4) first to pH 8.3 using phenolpthalein as an indicator and then further to pH between 4.2 and 5.4 with methyl orange or mixed indicator. In first case the value is called as phenolpthalein alkalinity (PA) and in second case, it is total alkalinity (TA) values of carbonates, bicarbonates and hydroxyl ions can be computed from two types alkalinity.

Reagents

(a) Hydrochloric acid, 0.1N

Dilute 12N concentrated HCl to 12 times to prepare 1.0N HCl. Dilute it further to make 0.1 HCl. (100-1000 ml). Standardized it against sodium carbonate solution .

(b) Methyl Orange indicator : (0.05%)

Dissolve 0.5 gm of methyl orange in 100 ml of distilled water.

(c) Phenolpthalein indicator :

Dissolve 0.5 gm. of phenolpthalein in 50 ml. of 95% ethanol and add 50ml. of distilled water. Add .05N. CO_2 free NaOH solution dropwise until the solution turns faintly pink.

Procedure

Take 100 ml of sample in a conical flask and add 2 drops of phenolpthalein indicator. If the solution remains colorless PA=0 and total alkalinity is determined as described in step 4. If the color changes to pink after addition of phenolpthalein, titrate it with 0.1 N HCl untill the color disappears at end point. This is phenolpthalein alkalinity (PA). Now add 2-3 drops of methyl orange to same sample and continue the titration further until the yellow color changes to pink at end point. This is total alkalinity (TA).

Calculations

$$\text{Total Alkalinity (ppm)} = \frac{\text{ml of HCl used with phenolpthalein and methyl orange} \times \text{N} \times 1000 \times 50}{\text{ml of sample}}$$

(12) *Hardness : (EDTA method)*

Hardness is generally caused by the calcium and magnesium ions present in water, polyvalent ions of some other metals like strontium, iron, aluminium, zinc and manganese etc. are also capable of precipitating the soap and thus contributing to hardness. However, the concentration of these ions is very low in natural waters. Therefore hardness is generally measured as concentration of only calcium and manganese as $CaCO_3$ which are far higher in quantities over other hardness producing ions.

Calcium and manganese form a complex of wine red color with erichrome-black-T at of 10±0.1 EDTA has get a stronger affinity towards Ca^{++} and Mg^{++} and therefore, by addition of EDTA, the former complex is broken down and a new complex of blue color is formed.

Reagents :

(a) EDTA solution

Dissolve 3.723 m of di-sodium salt of EDTA in distilled water to prepare 1 litre of solution.

(b) Buffer solution

Dissolve 16.9 gm of NH_4Cl in 143 ml of concentrated NH_4OH. Dissolve 1.179 gm of di-sodium EDTA and 0.780 gm of $MgSO_4$ in distilled water. Now mix both and dilute 250 ml distilled water.

Erichrome black-T indicator

Dissolve 0.5 gm of erichrome black-T dye in 100 ml of 80% ethyl alchol.

Procedure

Take 100 ml of sample in conical flask. If the sample is having higher calcium, take a smaller volume and dilute to 50 ml–Add 2 ml of buffer solution, add 2-3 drops of erichrome black-T indicator, the solution turns wine red. Titrate the contents against EDTA solution. At the end point color changes from wine red to blue.

Calculation

$$\text{Hardness (as ppm./l } CaCO_3) = \frac{\text{ml of EDTA} \times 1000}{\text{ml of sample}}$$

(13) *Calcium*

Many indicators such as ammonium etc. form a complex with only calcium but not with magnesium together. EDTA is having a higher affinity towards calcium, the former complex is broken down and a new complex is formed. However EDTA is a property to combine with both Ca^{++} and Mg^{++} therefore magnesium is largely precipitated on its hydroxide at sufficiently higher pH.

Reagents

(a) EDTA Solution

Dissolve 3.723 gm. of di-sodium salt of EDTA in distilled water to prepare 1 litre of solution. Store it into polythene or pyrex bottle.

(b) Sodium Hydroxide Solution (8%)

Dissolve 8 gm of sodium hydroxide in 100 ml of distilled water to prepare solution.

(c) Murexide Indicator

Mix 0.2 gm of ammonium purpurate with 100 gm. NaCl and grind thoroughly.

Procedure

Take 100 ml of sample in conical flask. If the sample is having higher alkalinity, use small volume dilute to 50 ml Add 2.0 ml of NaOH solution in sample.

Add 100-200 mg of murexide indicator, a pink color develops. Titrate against EDTA solution until the pink color changes to purple.

Calculations

$$\text{Ca (ppm.)} = \frac{X \times 400.8}{\text{ml of sample}}$$

where,

X = volume of EDTA used

(14) *Magnesium*

Calcium and magnesium form a complex of wine red color with erichrome black-T at pH 10.0. The EDTA has got a stronger affinity for Ca^{++} and Mg^{++} the former complex is broken down and a new complex of blue color is formed. The value of 'Mg' can be obtained by substracting the value of Ca from the total of Ca Mg.

Reagents

a- EDTA solution

b- Buffer solution

c- Erichrome Black-T

Procedure

Find out the volume of EDTA used in 'Ca' determination also find out the volume of EDTA used in hardness ($Ca^{++}Mg^{++}$) determination with same volume of sample as taken in Ca determination.

Calculation

$$\text{Mg (ppm.)} = \frac{(y\text{-}x) \times 400.8}{\text{volume of sample} \times 1.645}$$

y = EDTA used in hardness determination

x = EDTA used in calcium determination for the same volume of sample

(15) *Chloride*

Silver nitrate reacts with chloride to form a very slightly soluble white precipitate of AgCl at the end point when the Ca gets precipitated free silver ions get reacts with chromate to form a chromate of redish brown color.

Reagents

(a) Silver Nitrate Solution 0.2N

Dissolve 3.397 gm of dried $AgNO_3$ in distilled water to make 1 litre of solution and keep in dark bottle.

(b) Potassium Chromate Indicator 5%:

Dissolve 5gm of K_2CrO_4 in 100 ml of distilled water.

Procedure

Take 100 ml of sample in a conical flask and add 2 ml of K_2CrO_4 solution. Titrate the contents against .02N $AgNO_3$ until a persistent red tinge appears.

Calculations

$$\text{Cl (ppm.)} = \frac{\text{ml} \times \text{N of } AgNO_3 \times 1000 \times 35.5}{\text{ml. of sample}}$$

(16) ***Sodium : (Flame Photometric Method)***

A characteristic light is produced due to excitation of electrons when the sample is sprayed into the flame. The intensity of this characteristic radiation is proportional to the concentration of sodium and can be read at 589NM by using suitable filter devices.

The standard above is a linear one at the lower concentration of sodium, however at higher concentration it has got a tendency to level off potassium and calcium may be interfere with the determination of sodium if the former is more than 5 times in ratio to sodium, and the latter 10 times. Magnesium will interfere only when its ratio 10 sodium increases much higher than 100 is rare occurrence. For determination of Na, the sample should be stored in polythene bottles to check the leaching of sodium from the glass.

Reagents

(a) Stock Sodium Solution

Dissolve 2.542 gm NaCl dried at 140°C in distilled water to make a 1 litre of solution.

(b) Intermediate Sodium Solution (100mg/1Na)

Dilute the stock Na solution 10 times (100 >1000 ml)

(c) Standard Sodium Solution

Dilute the intermediate solution further 10 times (100 >1000 ml)

Procedure

For non-polluted samples as where only dissolved sodium is to be estimated, filter the sample through filter paper to remove any suspended matter which otherwise may along the capillary of the instrument, find out the concentration of Na using flame photometer, follow the instructions provided by manufacturer for use, because of the difference in the make and models of differences instruments must of instruments, require preparation of calibration curves for determination of sodium depending on the initial concentration of Na in the sample. For accurate results if is better to use smaller ranges and if sample is having more Na if can be dilute to come in the range of determination.

Calculations

Na (ppm.) = Na in diluted aliquot × dilution factor

(17) ***Potassium : (Flame Photometric Method)***

Like sodium, potassium can also be determined accurately by flame photometer. The characteristic radiation for potassium is 768NM, the intensity of which can be read on a scale by using a filter for this wave-length.

Reagents

(a) Stock Potassium Solution (1000 ppm K)

Dissolve 1.907 gm of KCl dried at 110°C in distilled water to prepare 1 litre of solution.

(b) Intermediate Potassium Solution (100 mg/1K)

Dilute the stock solution 10 times (100>1000 ml)

(c) Standard Potassium Solution

Dilute the intermediate solution 10 times (100-1000 ml)

Procedure

Follow the same procedure as described for determination of sodium except that to use filter for potassium at 763 NM wave-length instead of Na filter.

Calculations

K (ppm.) = (mg/lK in diluted aliquot) × dilution factor.

(18) *Chemical Oxygen Demand: (COD)*

Chemical oxygen demand is the measure of oxygen consumed during the oxidation of the oxidizable organic matter by a strong oxidizing agent. Potassium dichromate in the presence of sulphuric acid is generally used as an oxidizing agent in determination of COD.

The sample is refluxed with $K_2Cr_2O_7$ and H_2SO_4 in presence of mercuric sulphate to neutralize the effect of chlorides, and silver-sulphates.

The excess of potassium dichromate is titrated against ferrous ammonium sulphate using ferroin as an indicator. The amount of $K_2Cr_2O_7$ used is proportional to oxidizable organic matter present in sample.

Reagents

(a) Potassium Dichromate Solution, 0.25N

Dissolve 12.259 gm of dried A.R grade $K_2Cr_2O_7$ in distilled water to make 1 litre of solution.

(b) Ferrous Amonium Sulphate, 0.1N

Dissolve 39.2 gm of $Fe(NH_4)_2(SO_4)_2.6H_2O$ in water adding 20 ml conc. H_2SO_4 to make 1 liter of solution standardize this solution with $K_2Cr_2O_7$.

For standardization, dilute 10.0 ml of $K_2Cr_2O_7$ to about 100 ml, add 30 ml conc. H_2SO_4, and titrate with ferrous ammonium sulphate using ferroin as an indicator.

(c) Ferrous Ammonium Sulphate, 01N

Dissolved 0.1 N ammonium sulphate to 10 times (100 > 100 ml)

(d) Ferroin Indicator

Dilute 1.485 gm. of 1.10 phenonthraline and 0.695 gm. of ferrous sulphate ($FeSO_4.7H_2O$) in distilled water to make 100 ml of solution.

(e) Sulphuric Acid

H_2SO_4 conc. (Sp.gr 1.84)

(f) Mercuric Sulphate

$HgSO_4$, Solid

(g) Silver Sulphate

Ag_2SO_4, Solid

Procedure

Take 20 ml. of sample in a 250-500 ml.COD flask with a ground joint for Liebig reflux condensor.

If the sample is expected to have COD more than 50 ppm, add 10 ml of 0.25 N potassium dichromate solution (in case COD expected less than 50 ppm 10 ml of 0.25N $K_2Cr_2O_7$). Add a pinch of Ag_2SO_4 and $HgSO_4$. If the sample contains chlorides higher amount $HgSO_4$ is added in the rate of to the chlorides, COD cannot be determined accurately if the sample contains chloride more than 2000mg./l, add 30ml of sulphuric acid, reflex at least for 2 hours on a hotplate. Remove the flask cool and add distilled water to make the final volume to about 140ml. add 2-3 drops of ferroin indicator, mix thoroughly and titrate with 0.1 N ferrous ammonium sulphate.

Run a blank with distilled water using same quantities of chemicals.

Calculation

$$\text{COD mg}/\text{l} = \frac{(\text{b-a}) \times \text{N ferrous ammonium sulphate} \times 1000 \times 8}{\text{ml of sample}}$$

where,

'a'= ml of titrant with sample
'b' = ml of titrant with blank

Heavy Metals

Heavy metals of water samples of Padali-Gujjar pond were analysed by **Atomic Absorption Spectro photometer** (A.A.S.) **Perkin Elmer Model No. 3110.** During pre-autumn, autumn and post-autumn seasons during 1998, all three sampling sectors.

Biological Parameters

Collection of Plankton

The plankton of the pond were collected by fine plankton net. The known amount of water was filtered , and concentrated in the desired quantity. 1 ml of the concentrated was taken and placed in a sedge wick rafter counting cell. Counting of organisms was done by applying following formula :

$$N = \frac{(A \times 1000)C}{l}$$

where,

N = number of plankton / ml of water.

A = average no. of plankton in one small counting chamber of sedge wick rafter counting cell.

C = ml of plankton concentrate.

L = volume of original water filtered in liter.

The plankton were identified with the help of APHA (1980), Palmer (1980) and Needham and Needham (1972).

Standard Plate Count (SPC)

In solid medium, counting of organism depends on the fact that living cells will proceed to multiply and in times will produce sufficient progeny to form a colony visible to naked eyes. Since bacteria occurring in water as single cells, pairs, groups, chains or even dense clumps, not every individual living cells will develop into a separate colony on incubation. Therefore a number of colonies appearing on a plate does not necessarily represent the total number of organisms present in test volume. The results are expressed as the number of colonies/ml.

Apparatus and Material

(1) Incubator with arrangement at stable 37°C
(2) Petridishes, diameter 10cm
(3) Conical flask 250ml
(4) Thermometer
(5) Bacteriological pipettes
(6) Gas burner for maintaining the aseptic conditions
(7) Colony counter
(8) Water bath
(9) Culture tubes (borosil)
(10) Nutrient Agar Medium (NAM)

Beef extract	3.0 gm
Peptone	5.0 gm
Glucose	1.0 gm
Agar-agar	15.0 gm
Distilled water	1.0 liter
pH	7.0 ± 0.1

Preparation of Serial Dilution

Prepare 4 test tubes each with 9.0 ml distilled water sterilize all the tubes after plugging with cotton. Now shake the sample vigourously and transfer. 1 ml liquior from the first tube, transfer again 1.0 ml liquid from the first tube to second and so on. Use separate sterilized pipettes for each transfer. The dilution will be 1 : 10, 1 : 100, 1 : 1000 and 1: 10000...

Procedure

Select dilution according to the expected SPC .

Prepare Nutrient Agar Medium place the medium after melting, in a water bath maintained at 44-46°C for 1 hour. Transfer 1 ml of each 1, 1:10, 1:100, 1:10000 dilution in aseptic condition to already marked plates (sterilized). After delivering the sample, touch the tip of the pipette to a dry spot in the plates.

Now pour 12-15 ml liquified medium of the temp 44-46°C to these petriplates by gently opening the plates slightly and pouring the sample in petriplates invert the plates and keep for incubation at 37°C $\pm$ 0.5°C for 48 $\pm$ 3 hrs. Count the colonies after incubation in a colony counter, record the results and calculates the SPC/ml of water as follows.

$$\text{SPC/ml} = \frac{\text{Colonies counted}}{\text{Dilution factor}}$$

Most Probable Number (MPN)

The Coliform group comprises all aerobic and facultative anaerobic, Gram negative, none spore forming , rodshaped bacteria that ferment lactose with gas formation within 48 hrs at 35°C.

It includes presumptive test, confirmatory test and completed test.

Apparatus and Materials

(1) Fermentation tubes 25ml, 50ml
(2) Inoculation loops
(3) Durham vials
(4) Incubator
(5) Pertridishes
(6) Conical flasks
(7) Bacteriological pipettes
(8) Spirit lamp
(9) Autoclave
(10) Thermometer

(a) Presumptive Test

"Mac-Conkey's Broth" is required for this test.

The composition of broth is as below:

Bile salt	5.0 gm
Peptone	20.0 gm
Lactose	10.0 gm
NaCI	5.0 gm
Bromocrisol purple	1 ml
Distilled water	1 litre
Final pH	7.4 - 7.5

Procedure

Prepare several dilution of the sample up to 1:100 are required. Take 5 fermentation tubes for each dilution. Pour these tubes with 9.0 ml broth and put one durham vial to each tube in inverted condition. Now plug the mouth of all the tubes with cotton plugs. Now sterilize all these tubes in autoclave . Shake all the dilution rigorously before moving the sample to the fermentation tubes. Separate sterilized pipettes should be used for different dilutions as well as samples, shake fermentation tubes gently. Now incubate these tubes for 48 hrs at 35°C-37°C temperature for gas production. Record the tubes as positive (±), showing the gas production in durham viales. Even a tiny bubble should be recorded as positive, the tubes showing positive results are now subjected to the confirmatory test. Incubate negative tubes for further 24hrs to confirm the test.

(b) Confirmatory test

For this test "Brilliant Green Lactose Broth" (BGLB) is required the composition is as given below :

Peptone	10.0 gm
Lactose	10.0 gm
Oxgall	20.0 gm
Brilliant green	1.0 litre
pH	7.2

Procedure

Prepare fermentation tubes with 10 ml BGLB medium and durham's vial. The number of tubes should be equal to the all positive tubes in the presumptive test. Shake positive tubes gently. Transfer 1 loop full of medium to the tubes having BGLB broth with the help of inoculation niddle. Incubate these tubes for 48 hrs at 30-37°C.

Record the tubes as positive which are showing gas production. Put the number of positive tubes in the MPN table and then in the formula to get the MPN/100 ml. To confirm this test try for completed test.

(c) Completed Test

For this test "Eosin Methylene Blue agar" (EMB) medium is required. The composition of EMB is as given below:

Peptone	10.0 gm
Lactose	10.0 gm
K_2HPO_4	2.0 gm
Eosin y	0.4 gm
Methylene blue	0.065 gm
Agar-agar	15.0 gm
Distilled water	1.0 litre
pH	7.1

Procedure

Prepare EMB agar pertriplates. The number of the positive tubes, of the confirmatory test streak the inoculum from the positive BGLB tubes on these plates and incubate for 24 hrs at 37°C examine the plate for bacterial colony appearance.

Fish

Fish Collection and Preservation

The Padali-Gujjar Pond is a natural pond at Roorkee, is being utilized for the commercial fish production since 20-25 years.

The composite fish culture practice adopted by a group of fish farmers (local fish farmer society). The quantity of the seeds stocking material were stocked in the ratio of

Fig. 4.1 Operation of Gillnet during harvesting

Fig. 4.2 : Operation of Gillnet during harvesting

15:25:15:15:10:10:10, namely *Catla-catla, Labeo-rohita, Cirrihinas-mrigala, Hypophthalmichthys-molitrix* (Silver carp.), *Aristichthys-nobalis* (Big head), *Cyprinus-carpio* (Common carp.), *Ctenopharyngodon-idella* (Grass carp.) in the month of July and August.

No supplementry feeds were given for the fish growth, so the culturable fishes are depends upon natural productivity of the pond.

At the time of harvesting of fish production was estimated by operating deep water gill nets of different dimensions ranged 16.5 to 10 to 75.0 ×10 with a mesh size ranging 50-120 mm knot to knot (Figs. 4.1 and 4.2)

The catch composition and Catch Per Unit Effort was estimated by following method:

$$\text{Catch Per Unit Effort (CPUE)} = \frac{\text{Total weight of catch}}{\text{No. of persons} \times \text{hours of fishing}}$$

The fish caught were collected and preserved in 10% formalein for identification.

They were identified according to Gunther *et al.* (1959), Day (1879) and Jayaram (1987).

STATISTICAL METHOD

Standard Deviation

Standard deviation was calculated to find out the fluctuation from the average mean value by the following formula:

$$\sigma = \sqrt{\Sigma\ (x\text{-}x)^2 / N}$$

CHAPTER -5

RESULTS

(A) Physico-chemical Parameters

Physico-chemical parameters imparts a major role to determine the quality of natural waters. The limnological parameters of the "Padali-Gujjar Pond" studied at Roorkee.

The observations made during Pre-autumn, Autumn and Post-autumn seasons during 1998 for the following parameters. Observation of physical and chemical parameters are showed in (Tables 5.1 to 5.6 and Figs. 5.1 to 5.16)

(A) Physical Parameters

Water temperature	(°C)
Total solids (T.S.)	(ppm.)
Total dissolved solids (T.D.S.)	(ppm.)
Total suspended solids (T.S.S.)	(ppm.)
Turbidity	(J.T.U.)
Conductivity	(micro mho)

(B) Chemical Parameters

pH
Dissolved oxygen (ppm.)
Biological oxygen demand (ppm.)
Chemical oxygen demand (ppm.)
Free carbon-di-oxide (ppm.)
Alkalinity (ppm.)
Chlorides (ppm.)
Hardness (ppm.)
Calcium (ppm.)
Magnesium (ppm.)
Sodium (ppm.)
Potassium (ppm.)

(C) Heavy Metals

Zinc (Zn) (ppm.)
Iron (Fe) (ppm.)
Manganese (Mn) (ppm.)
Cobalt (Co) (ppm.)

SAMPLING SECTOR-I

(A) Physical parameters

Water temperature

The water temperature of Padali-Gujjar Pond at this sampling sector was observed maximum in the Pre-autumn (31.7°C±0.56) and minimum in the Post-autumn season (23.8°C ±3.3). The average of water temperature was recorded for the Padali-Gujjar Pond was 28.0°C±3.23.

Total Solids (T.S.)

The maximum value of total solids in pond water was recorded in Pre-autumn season (325.24ppm±7.03) and minimum in the Post-autumn season (259.24ppm±7.05). The average value of total solids was 293.4ppm±26.95.

Total Dissolved Solids

The total dissolved solids in pond water as sampling sector first was observed maximum in pre-autumn (189.82ppm±14.65) and minimum in post-autumn (149.71ppm±5.27). The average of total dissolved solids in water of sampling sector first was 164.1ppm±18.19.

Total Suspended Solids

Total suspended solids was maximum in pre-autumn (55.04 ppm + 23.94) and minimum in post-autumn (30.50ppm±3.04). The average of total suspended solids in this sampling sector was 45.50ppm±10.80.

Turbidity

The turbidity of water was calculated by **Jackson's candle turbidity meter**. The maximum value of turbidity at Padali-Gujjar Pond at this sampling sector was observed 133.75 J.T.U. ± 65.40 during pre-autumn and minimum in post-autumn 19.7 J.T.U. ± 2.47. The average of turbidity was observed 81.75 J.T.U.± 47.09.

Conductivity

The conductivity of water samples was calculated by **digital conductivity** meter. The maximum value of conductivity was calculated in pre-autumn season (194 micro mho±36.76) and minimum in post-autumn (165 micro mho±21.21). The average of conductivity in water of sampling sector first was recorded 183.3 micro mho±12.92.

(B) Chemical Parameters

pH

During the period of study the pH value of pond water at sampling sector first was highest in pre-autumn (9.37±0.48) and lowest value in post-autumn season was 9.05±0.17. The average of pH was recorded 9.10±0.19.

Dissolved Oxygen

The value of dissolved oxygen of the Padali-Gujjar Pond was observed minimum in pre-autumn season (11.05ppm±0.96) and maximum in post-autumn season (11.53ppm ± 4.72). The average of dissolved oxygen for this sector was 10.99ppm ± 0.05.

Biological Oxygen Demand

The BOD of pond water in this sector was recorded maximum in pre-autumn season (3.10ppm±0.19) and minimum in post-autumn (2.81±0.09). The average of BOD was observed 2.95ppm±0.35.

Chemical Oxygen Demand

During the period of study, the COD of water at this sector was calculated maximum in pre-autumn season (8.30ppm±0.88) and minimum in post-autumn season (6.71ppm ± 0.04).The average value at sector first was calculated 7.40ppm ±1.01.

Free carbon-di-oxide

Free carbon-di-oxide in water of pond at sampling sector first was observed nil during pre-autumn season and minimum in post-autumn season 1.96ppm ± 2.1 and maximum in post-autumn season 2.0ppm ± 0.14 and average of free CO_2 was observed 1.32ppm ± 0.93.

Alkalinity

The alkalinity of water at this sector was recorded to be minimum in post-autumn season (70.05ppm ± 1.80) and maximum in pre-autumn season (70.5ppm ± 3.12). The average value was recorded 69.46ppm ±1.46.

Chloride

The minimum value of chloride concentration at sampling sector first was recorded 61.20ppm ±1.15 in post-autumn season and maximum 71.7ppm ± 1.24 in pre-autumn. The average of chloride concentration was recorded 65.91ppm ± 4.35.

Hardness

The total hardness of Padali-Gujjar Pond at this sector was recorded minimum at this sector 61.27ppm ± 1.15 in post-autumn season and maximum in pre-autumn 75.80ppm ± 0.82. The average value was recorded 68.60ppm ± 5.93.

Calcium

The calcium concentration of the pond water was observed to be minimum 65.80 ppm ± 3.63 in post-autumn and maximum in pre-autumn 68.20 ppm ± 0.67 and average value was 65.75ppm ± 2.02.

Magnesium

The maximum concentration of the magnesium was recorded at this sector in pre-autumn season (7.40 ppm ± 0.06) and minimum (5.30 ppm ± 0.67) in autumn season. The average value of magnesium concentration was recorded 6.26 ppm ± 0.96.

Sodium

Sodium concentration in pond water was recorded maximum 4.0 ppm ± 0.59 during pre-autumn seasons and minimum 2.50 ppm ± 0.46 during autumn season and the average concentration was 3.16 ppm ± 0.62.

Potassium

Minimum value of potassium was observed 1.52ppm ±1.44 in post-autumn season and maximum 5.60ppm ± 1.37 in autumn season. The average value was recorded 3.63 ±1.63.

SAMPLING SECTOR : II

Physical Parameters

Water Temperature

The temperature of water at Padali-Gujjar Pond at sampling sector second was observed minimum (25.2°C ± 4.10) in post-autumn season and maximum in pre-autumn season (30°C ± 0.28). The average of water temperature recorded 27.80°C ± 1.51.

Total Solids (T.S.)

The total solids was observed maximum in pre-autumn (310.60ppm ± 13.73) and minimum in post-autumn (233.40ppm ± 9.87). The average of total solids in this sector water was calculated 263.20ppm ± 20.16.

Total Dissolved Solids

The minimum value of total dissolved solids in pond water was recorded (112.30ppm ± 2.05) in post-autumn season and maximum in pre-autumn season (200.3ppm ± 0.46). The average of total dissolved solids was recorded 150.40ppm ± 36.87.

Total Suspended Solids

The value of total suspended solids in sampling sector second water was observed maximum (121.20ppm ± 7.82) in post-autumn season and minimum in pre-autumn season (110.20ppm ± 13.26). The average was observed 11 4.30ppm ± 4.73.

Turbidity

The turbidity of Padali Gujjar Pond was calculated by **Jackson's candle turbidity meter**. The maximum value of turbidity was recorded 132.50 J.T.U. ± 24.74 in pre-autumn season and 22.50 J.T.U. ± 3.53 minimum is post-autumn season. The average of turbidity in this sector was recorded 82.16 J.T.U. ± 45.36.

Conductivity

The conductivity of water was calculated by digital conductivity meter. The minimum value of conductivity was calculated 153 micro mho ± 4.24 in post-autumn season and maximum value was 185 micro mho ± 35.35 in pre-autumn. The average of conductivity in water was calculated 174.30 micro mho ± 15.08.

Chemical Parameter

pH

There was little variations in pH. During the period of study the maximum value was observed 8.62 ± 0.82 in post-autumn season and minimum in autumn season 8.02 ± 0.28. The average of pH was observed in this sector 8.22 ± 0.28.

Dissolved Oxygen

The value of dissolved oxygen was minimum (9.05 ppm. $\pm$ 0.60) during pre-autumn season and maximum(10.91 ppm. $\pm$ 0.79) in post-autumn season. The average of dissolved oxygen in this sector was observed 9.69 ppm $\pm$ 0.85.

Biochemical Oxygen Demand

The BOD of pond water in this sector was recorded maximum (2.98ppm.$\pm$ 0.12) in pre-autumn seasons and minimum (2.08ppm. $\pm$ 0.05) during post-autumn season. The average BOD was recorded 2.64ppm. $\pm$ 0.03.

Chemical Oxygen Demand.

During the period of study. The COD of Padali Gujjar pond of this sector was calculated maximum (8.09±0.38) in pre-autumn season and minimum (6.04ppm. $\pm$ 0.04) in post-autumn season. The average of COD in pond water of this sector was calculated 7.08ppm. $\pm$ 0.11.

Free Carbon di-oxide

Free carbon-di-oxide in water at this spot was observed maximum (2.3ppm. $\pm$ 0.42) in pre-autumn season and minimum (1.35 ppm. $\pm$ 0.35) in post-autumn season. The average free CO_2 was calculated 1.95mg / 1. $\pm$ 0.42.

Alkalinity

The alkalinity of water at this sector was recorded to be minimum 68.7ppm 6.94 during pre-autumn season and maximum 70.70ppm $\pm$ 2.99 in post-autumn season. The average alkalinity was observed 69.10ppm. $\pm$ 1.12.

Chlorides

The maximum value of chlorides concentration was recorded (71.22ppm. $\pm$ 132) during pre-autumn season and minimum (65.35ppm. $\pm$ 4.78) in autumn season. The average of chloride concentration was recorded 67.99ppm. $\pm$ 2.42.

Hardness

The total hardness of Padali-Gujjar pond water at this sector was calculated maximum (77.10ppm+0.34) during pre-autumn season and minimum (71.21ppm. $\pm$ 4.20) during autumn season. The average value was calculated 73.24ppm. $\pm$ 2.65.

Calcium

The calcium concentration of the pond water was observed, maximum (68.72ppm. $\pm$ 0.56) in pre-autumn season and minimum (64.90ppm. $\pm$ 2.17) during autumn season. The average value was recorded 66.68ppm. $\pm$ 1.56.

Magnesium

Maximum concentration of magnesium was recorded at this sector in pre-autumn season (8.39ppm. $\pm$ 0.21), and minimum (5.25ppm. $\pm$ 0.01) in post-autumn season. The average value of magnesium concentration was recorded 6.65ppm. $\pm$ 2.22.

Sodium

Maximum value of sodium in pond water was observed (4.80 ppm ± 0.82) in pre-autumn season and minimum (3.05 ppm ± 0.41) in post-autumn season and average was calculated 3.63 ppm ± 0.82.

Potassium

During the study period of sampling sector second the concentration of potassium was found maximum (4.23 ppm ± 0.60) in autumn season and minimum (2.0ppm ± 0.97) in post-autumn season. The average concentration of potassium was observed 3.38ppm ± 0.98.

SAMPLING SECTOR - III

Physical Parameters

Water Temperature

The water temperature of Padali-Gujjar pond at sampling sector third was observed maximum, (30.4°C ± 1.48) in pre-autumn and minimum (23.60°C ± 5.09) in post-autumn season. The average of pond water at this sector was observed 27.7°C ± 4.34.

Total Solids

Maximum value of total solids was observed (309.40ppm. ± 12.91) during pre-autumn season and minimum (257.70ppm. ± 20.46) in post-autumn season. The average was observed in this sector 283.80ppm. ± 21.11.

Total Dissolved Solids

The value of total dissolved solids in this sector was observed minimum (226.75ppm.± 17.36) in post-autumn season, and maximum (254.50ppm.±11.73) in pre-autumn season. The average of total dissolved solids in pond water was observed 238.3ppm. ± 11.77.

Total Suspended Solids

Total suspended solids was maximum (142.95ppm. ± 4.70) in autumn season and minimum (109.53ppm. ± 6.98) in post-autumn season. The average was observed 129.30ppm.± 14.28.

Turbidity

The turbidity of pond water in sampling sector third was observed by **Jackson's candle tubidity meter**. The minimum value of turbidity at Padali-Gujjar pond in sampling sector third was observed (29.5J.T.U. ± 3.53) in post autumn and maximum(58.75J.T.U. ± 30.05) in pre-autumn. The average of turbidity was observed 47.7J.T.U. ± 12.99.

Conductivity

The conductivity of water was calculated by digital conductivity meter. The minimum value of conductivity was calculated (155micro mho ± 4.24) during post autumn season and maximum value (185 micro mho ± 43.84) during pre-autumn season. The average of conductivity in sampling sector third was calculated 173.66 micro mho ± 16.99.

Chemical Parameters

pH

There was little variations in pH at sampling sector third of Padali-Gujjar Pond. The minimum value was observed (7.69 ± 0.24) during pre-autumn season and maximum (7.98 ± 0.06) during autumn season. The average of pH was recorded 7.81 ± 012.

Dissolved Oxygen

The maximum value of dissolved oxygen in this sampling sector was calculated (9.86ppm. ± 038) in post-autumn season and minimum (7.94ppm. ± 0.58) in pre-autumn season. The average of dissolved oxygen was calculated 8.89ppm. ± 0.80.

Biochemical Oxygen Demand

Water BOD at Padali Gujjar Pond in sampling sector third was observed minimum (1.79ppm. ± 0.17) during post-autumn season and maximum (2.47pprn ± .0.57) during autumn season. The average value was found 2.18ppm. ± 0.20.

Chemical Oxygen Demand

The COD of pond water at this sector during period of study was calculated maximum (6.67ppm. ± 0.34) in pre-autumn season and minimum (5.50ppm. ± 0.10) in post-autumn season. The average of COD was observed 6.35 ppm. ± 0.35.

Free Carbon-di-oxide

During the period of study at this spot free CO_2 was observed minimum (2.25ppm. ± 0.07) in post-autumn season and maximum in (2.60ppm. ± 0.42) in pre-autumn season. The average of free CO_2 at this sector was observed 2.4ppm ± 0.08.

Alkalinity

The alkalinity of water was recorded to be maximum (81.41ppm ± 1.49) in pre-autumn season and minimum (78.87ppm ± 0.62) in autumn season. During period of study at Paddali-Gujjar Pond the average alkalinity was recorded 80.06ppm ± 1.03.

Chlorides

Maximum value of chloride concentration at this sector was calculated (35.70ppm ± 5.03) in pre-autumn season and minimum (34.78ppm ± 2.39) in post autumn season. The average was calculated 35.18ppm ± 0.29.

Hardness

The total hardness of pond water was observed maximum (76.60ppm ± 0.55) in pre-autumn season and minimum (69.31ppm ± 1.31) in post-autumn season. In sampling sector third the average value was observed 72.11 ppm ± 2.72.

Calcium

Maximum concentration of calcium was calculated (63.22ppm ± 0.05) during pre-autumn season and minimum (60.05ppm ± 3.28) during post-autumn season. The average was to be 61.24ppm ± 1.25.

Table 5.3 : Physical parameters of the water of *Padali-Gujjar Pond* during different seasons (1998) at sampling sector–II

S. No.	*Physical Parameters*	*Pre- Autumn*	*Autumn*	*Post-Autumn*	*Average*
1	Temperature (°C)	30 ± 0.28	28.30 ± 1.48	25.20 ± 4.10	27.80 ± 1.51
2	Total solids (ppm.)	310.60 ± 13.73	245.60 ± 21.30	233.40 ± 9.87	263.20 ± 20.16
3	Total dissolved solids (ppm.)	200.30 ± 0.46	138.70 ± 13.70	112.30 ± 2.05	150.40 ± 36.87
4	Total Suspended solids (ppm.)	110.20 ± 13.26	111.90 ± 14.80	121.20 ± 7.82	114.30 ± 4.73
5	Turbidity (J.T.U.)	132.50 ± 24.74	19.50 ± 89.80	22.50 ± 3.53	82.16 ± 45.36
6	Conductivity (micro mho)	185 ± 35.35	185 ± 7.07	153 ± 4.24	174.30 ± 15.08

± = Standard Deviation

Table 5.4 : Chemical parameters of the water of *Padali-Gujjar Pond* during different seasons (1998) at sampling sector-II

S. No.	*Physical Parameters*	*Pre- Autumn*	*Autumn*	*Post-Autumn*	*Average*
1	pH	8.03 ± 0.28	8.02 ± 0.21	8.62 ± 0.82	8.22 ± 0.28
2	Dissolve oxygen (ppm.)	9.05 ± 0.60	9.13 ± 0.05	10.91 ± 0.79	9.69 ± 0.85
3	B.O.D (ppm.)	2.98 ± 0.12	2.86 ± 0.11	2.08 ± 0.05	2.64 ± 0.03
4	C.O.D (ppm.)	8.09 ± 0.38	7.15 ± 0.04	6.04 ± 0.04	7.09 ± 0.11
5	Free carbon-di-oxide (ppm.)	2.3 ± 0.42	2.20 ± 0.14	1.35 ± 0.35	1.95 ± 0.42
6	Alkalinity (ppm.)	68.17 ± 6.94	68.45 ± 4.27	70.70 ± 2.99	69.10 ± 1.12
7	Chlorides (ppm.)	71.22 ± 1.32	65.35 ± 4.78	67.40 ± 5.76	67.99 ± 2.42
8	Hardness (ppm.)	77.10 ± 0.34	71.21 ± 4.20	71.42 ± 2.89	73.24 ± 2.65
9	Calcium "Ca" (ppm.)	68.72 ± 0.56	64.90 ± 2.17	66.42 ± 3.23	66.68 ± 1.56
10	Magnesium "Mg" (ppm.)	8.39 ± 0.21	6.31 ± 2.02	5.25 ± 0.01	6.65 ± 2.22
11	Sodium "Na" (ppm.)	4.80 ± 0.82	3.06 ± 0.40	3.05 ± 0.41	3.63 ± 0.82
12	Potassium "K" (ppm.)	3.92 ± 0.38	4.23 ± 0.60	2.0 ± 0.97	3.38 ± 0.98

± = Standard Deviation

Table 5.5 : Physical parameters of the water of *Padali-Gujjar Pond* during different seasons (1998) at sampling sector-III

S. No.	*Physical Parameters*	*Pre- Autumn*	*Autumn*	*Post-Autumn*	*Average*
1	Temperature (°C)	30.40 ± 1.48	29.30 ± 1.13	23.60 ± 5.09	27.70 ± 4.34
2	Total solids (ppm.)	309.40 ± 12.91	284.50 ± 8.73	257.70 ± 20.46	283.80 ± 21.11
3	Total dissolved solids (ppm.)	254.50 ± 11.73	233.80 ± 12.21	226.75 ± 17.36	238.30 ± 11.77
4	Total suspended solids(ppm.)	135.42 ± 21.69	142.95 ± 4.70	109.53 ± 6.98	129.30 ± 14.28
5	Turbidity (J.T.U.)	58.79 ± 30.05	55 ± 28.28	29.50 ± 3.53	47.70 ± 12.99
6	Conductivity (micro mho)	185 ± 43.84	181 ± 21.21	155 ± 4.24	173.66 ± 16.99

± = Standard Deviation

Table 5.6 : Chemical parameters of the water of *Padali-Gujjar Pond* during different seasons (1998) at sampling sector-III

S. No.	*Physical Parameters*	*Pre- Autumn*	*Autumn*	*Post-Autumn*	*Average*
1	pH	7.69 ± 0.24	7.98 ± 0.06	7.76 ± 0.11	7.81 ± 0.12
2	Dissolve oxygen (ppm.)	7.94 ± 0.58	8.89 ± 0.10	9.86 ± 0.38	8.89 ± 0.80
3	B.O.D (ppm.)	2.29 ± 0.14	2.47 ± 0.57	1.79 ± 0.17	2.18 ± 0.20
4	C.O.D (ppm.)	6.67 ± 0.34	6.89 ± 0.19	5.50 ± 0.10	6.35 ± 0.35
5	Free carbon-di-oxide (ppm.)	2.60 ± 0.42	2.45 ± 0.07	2.25 ± 0.07	2.40 ± 0.08
6	Alkalinity (ppm.)	81.41 ± 1.49	78.87 ± 0.62	79.91 ± 2.05	80.06 ± 1.03
7	Chlorides (ppm.)	35.70 ± 5.03	35.06 ± 0.43	34.78 ± 2.39	35.18 ± 0.29
8	Hardness (ppm.)	76.60 ± 0.55	69.31 ± 1.31	70.96 ± 0.38	72.11 ± 2.72
9	Calcium "Ca" (ppm.)	63.22 ± 0.05	60.05 ± 3.28	60.45 ± 0.88	61.24 ± 1.25
10	Magnesium "Mg" (ppm.)	12.85 ± 0.50	9.27 ± 1.97	10.51 ± 0.49	10.87 ± 1.45
11	Sodium "Na" (ppm.)	3.70 ± 0.63	2.89 ± 0.64	1.81 ± 0.70	2.80 ± 0.93
12	Potassium "K" (ppm.)	2.76 ± 0.04	3.71 ± 0.62	2.03 ± 0.56	2.83 ± 0.68

± = Standard Deviation

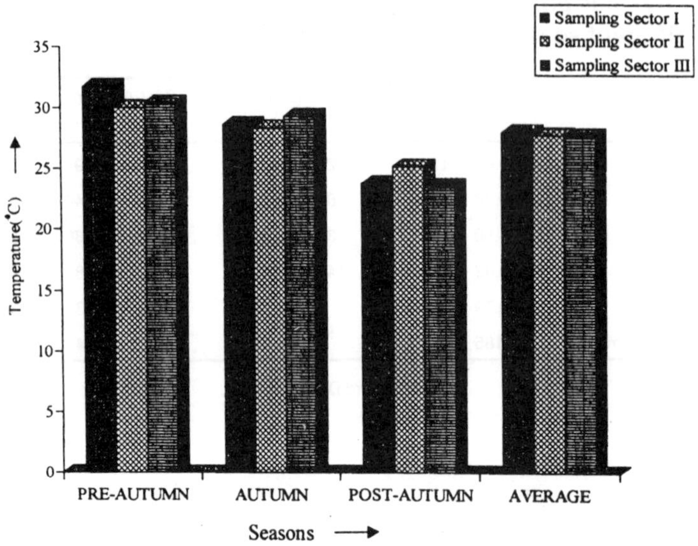

Fig. 5.1 : Seasonal Variation in Temperature (°C)

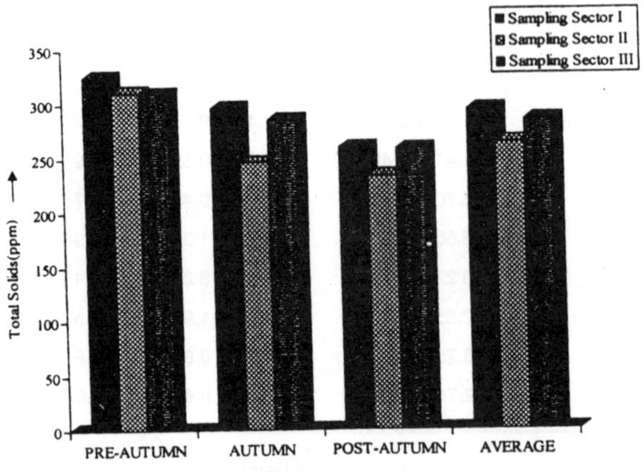

Fig. 5.2 : Seasonal Variation in Total Solids (ppm)

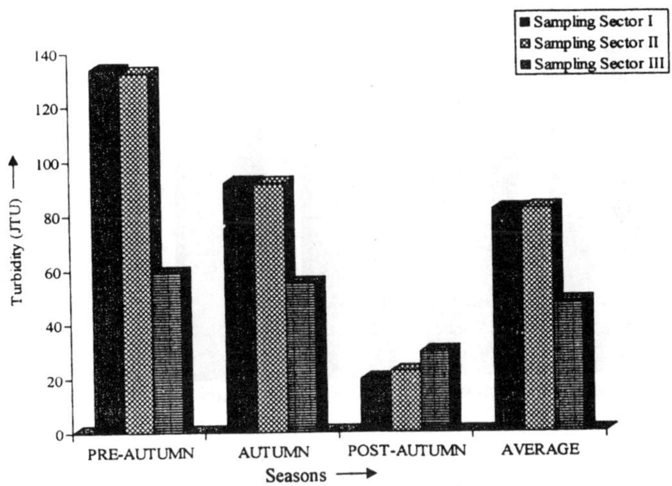

Fig. 5.3 : Seasonal Variation in Turbidity (JTU)

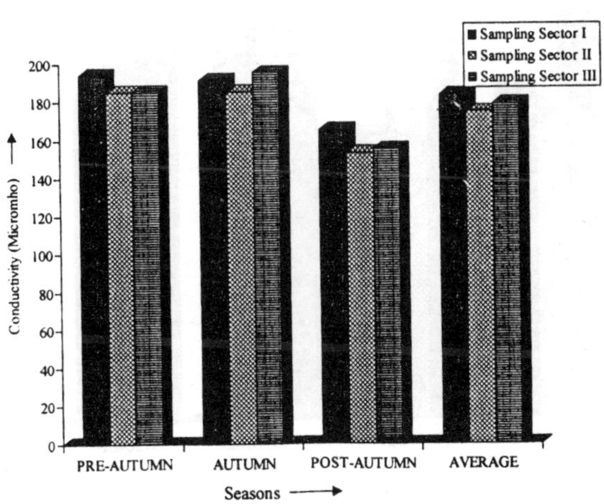

Fig. 5.4 : Seasonal Variation in Conductivity (Micromho)

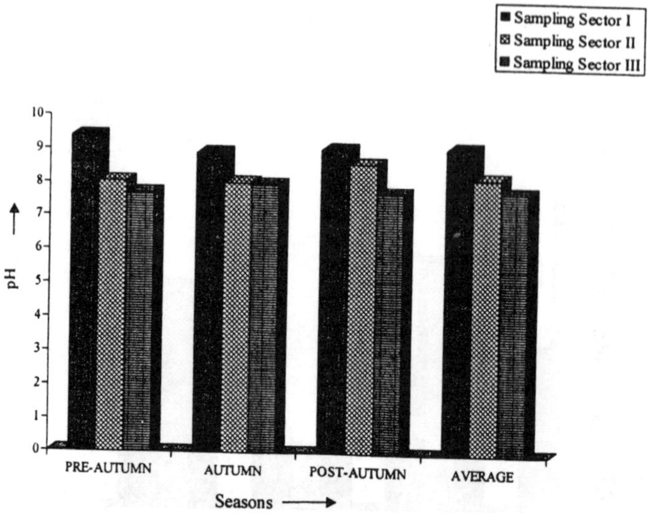

Fig. 5.5 : Seasonal Variation in pH

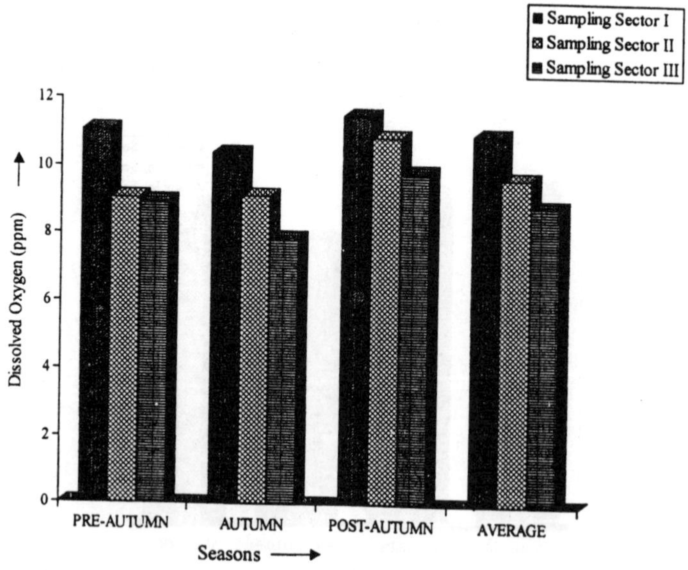

Fig. 5.6 : Seasonal Variation in Dissolved Oxygen (ppm)

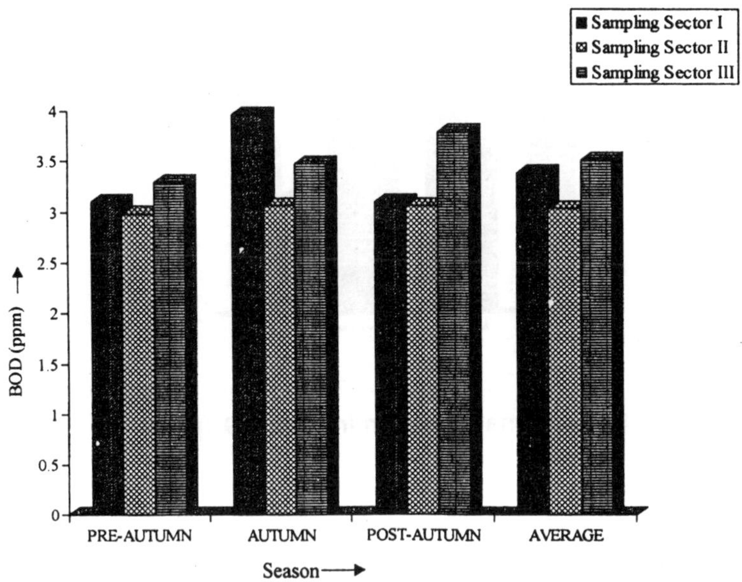

Fig. 5.7 : Seasonal Variation in BOD (ppm)

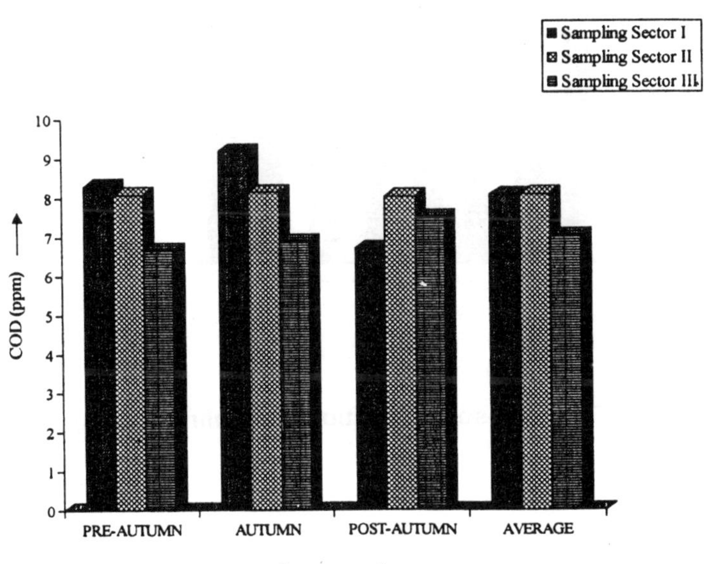

Fig. 5.8 : Seasonal Variation in COD (ppm)

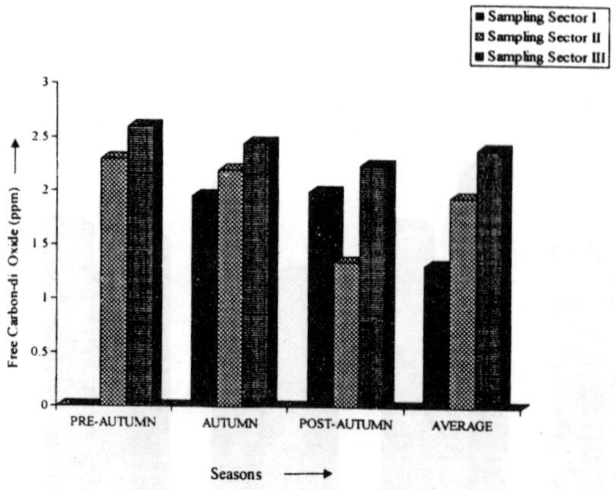

Fig. 5.9 : Seasonal Variation in Free CO_2 (ppm)

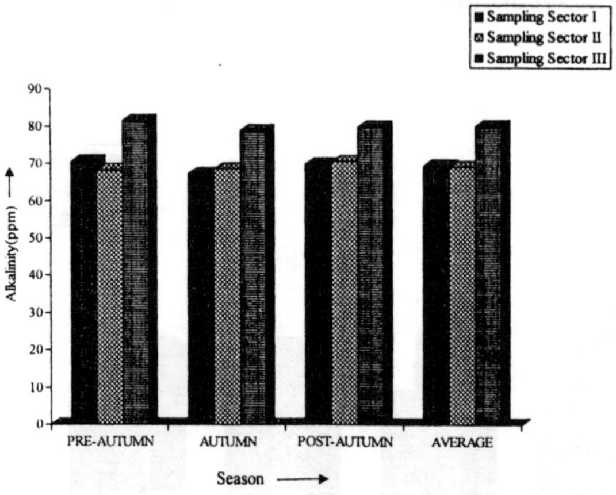

Fig. 5.10 : Seasonal Variation in Alkalinity (ppm)

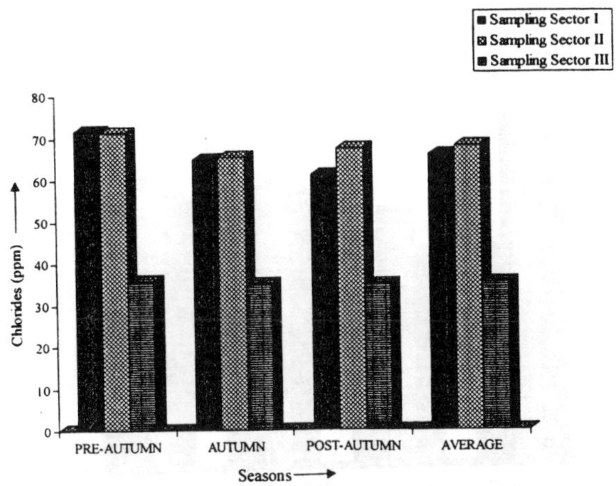

Fig. 5.11 : Seasonal Variation in Chlorides Concentration (ppm)

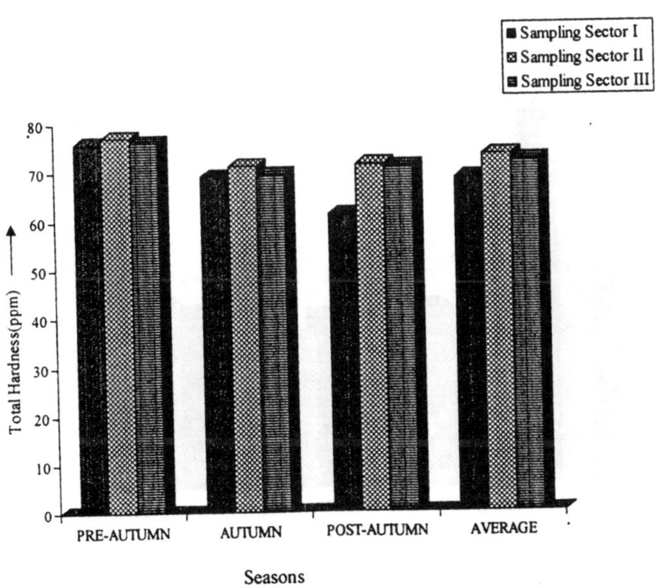

Fig. 5.12 : Seasonal Variation in Hardness (ppm)

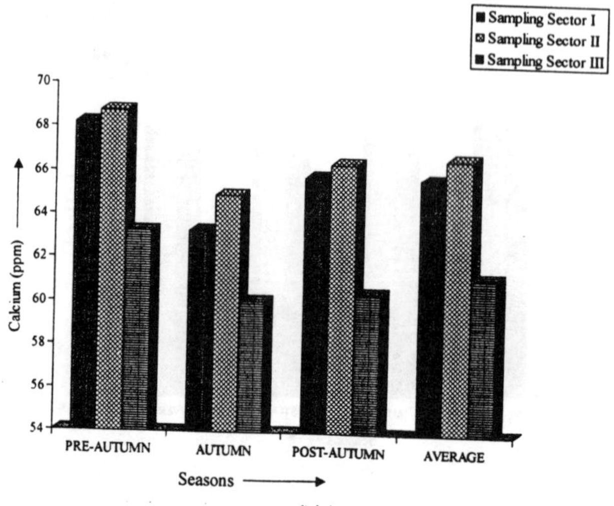

Fig. 5.13 : Seasonal Variation in Calcium Concentration (ppm)

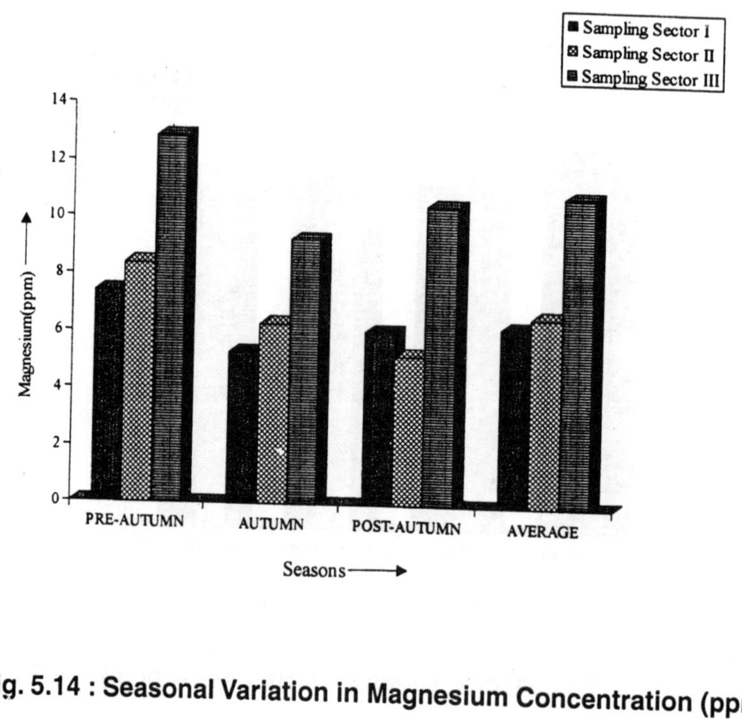

Fig. 5.14 : Seasonal Variation in Magnesium Concentration (ppm)

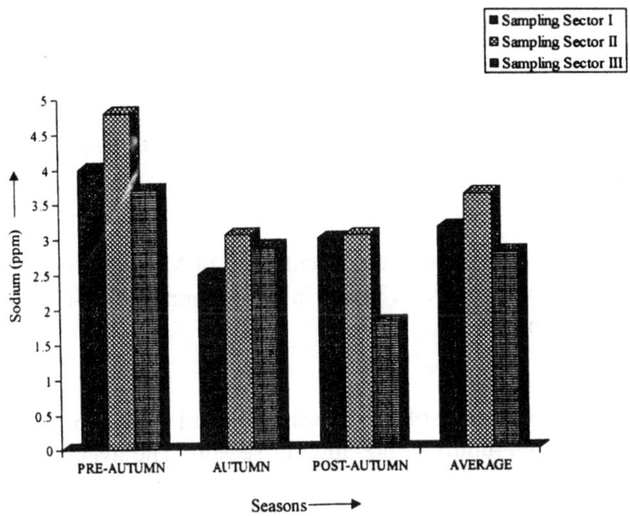

Fig. 5.15 : Seasonal Variation in Sodium Concentration (ppm)

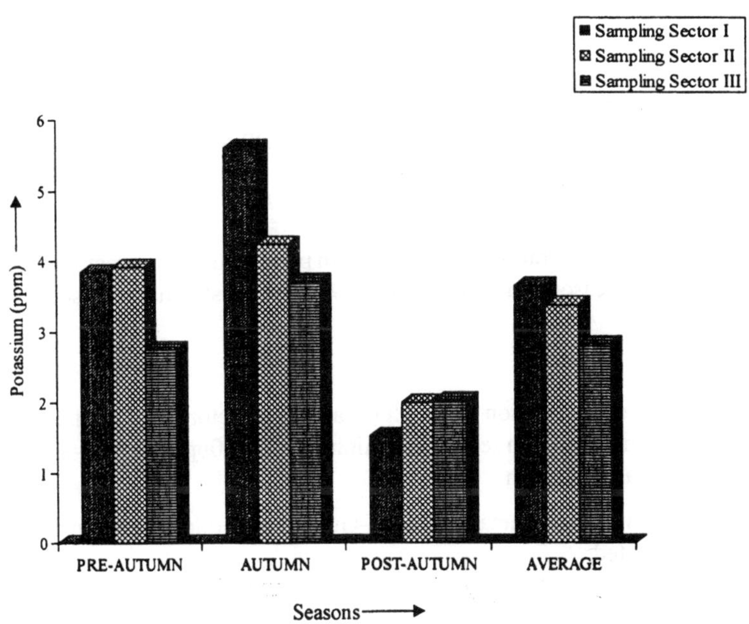

Fig. 5.16 : Seasonal Variation in Potassium Concentration (ppm)

(0.038ppm) in pre-autumn seasoson and minimum in autumn (0.032ppm) at sampling sector II. The average value was 0.035ppm.

Iron (Fe)

The maximum concentration of iron was observed 0.669ppm in pre autumn season and minimum 0.619ppm in autumn season in this sector. The average concentration was recorded 0.650ppm.

Manganese (Mn)

In sampling sector-II the minimum concentration of Mn was observed (0.135ppm) during autumn season and maximum 0.139ppm in pre-autumns. The average concentration in this sector was 0.136ppm.

Cobalt (CO)

The minimum concentration of cobalt was recorded (0.041ppm) in post-autumn season and maximum (0.044ppm) in pre-autumn season. The average was calculated 0.042ppm.

SAMPLING SECTOR-III

Zinc (Zn)

The concentration of Zinc in water of sampling sector-III was observed maximum (0.038ppm) in autumn season and minimum (0.035ppm) in post autumn season. The average value was 0.036ppm.

Iron (Fe)

Maximum concentration of iron in this sector was calculated (0.652ppm) during pre-autumn season and minimum (0.640ppm) during autumn season. The average in this sector was calculated 0.647 ppm during study period.

Manganese

The maximum concentration of manganese in pond water at this sector was calculated (0.143ppm) in autumn season and minimum (0.139ppm) in post-autumn season. The average value was 0.140ppm.

Cobalt (CO)

The maximum concentration of cobalt in sampling sector-III of Padali-Gujjar Pond was found (0.043ppm) in autumn season and minimum (0.040ppm) in post-autumn season. The average value was 0.041ppm.

During analysis of heavy metals of Padali-Gujjar Pond following heavy metals Cu, Cr, Pb, Cd, are, not detected.

(D) Biological Parameters

Studies on the biological parameters of Padali-Gujjar Pond at Roorkee include the study of plankton, standard plate count (S.P.C.) Most Probable Number (MPN) and fishes. Tables 5.10 to 5.12 and Figs. 5.21 to 5.26 are showing the distribution of phytoplankton and distribution of zooplankton respectively at different sampling sectors–I, II, and III.

Table 5.7 : Seasonal variation in Heavy Metals of water of *Padali-Gujjar Pond*, Roorkeee during different seasons (1998) at Sampling Sector-I

S.No.	*Heavy metals (ppm.)*	*Pre-autumn*	*Autumn*	*Post-autumn*	*Average*
1	Zinc (Zn)	0.037	0.036	0.034	0.035
2	Iron (Fe)	0.783	0.780	0.785	0.782
3	Manganese (Mn)	0.157	0.150	0.153	0.153
4	Cobalt (CO)	0.050	0.051	0.053	0.051

Table 5.8 : Seasonal variation in Heavy Metals of water of *Padali-Gujjar Pond*, Roorkee during different seasons (1998) at Sampling Sector-II

S.No.	*Heavy metals (ppm.)*	*Pre-autumn*	*Auturnn*	*Post-autumn*	*Average*
1	Zinc (Zn)	0.038	0.032	0.036	0.035
2	Iron (Fe)	0.669	0.619	0.664	0.650
3	Manganese (Mn)	0.139	0.135	0.136	0.136
4	Cobalt (CO)	0.044	0.042	0.041	0.042

Table 5.9 : Seasonal variation in Heavy Metals of water of *Padali Gujar Pond*, Roorkee during different seasons (1998) at Sampling Sector-III

S.No.	*Heavy metals (ppm.)*	*Pre-autumn*	*Autumn*	*Post-autumn*	*Average*
1	Zinc (Zn)	0.036	0.038	0.035	0.036
2	Iron (Fe)	0.652	0.640	0.650	0.647
3	Manganese (Mn)	0.140	0.143	0.139	0.140
4	Cobalt (CO)	0.042	0.043	0.040	0.041

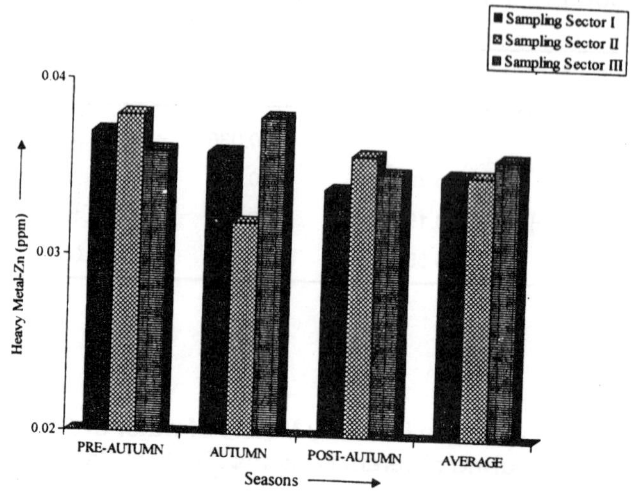

Fig. 5.17 : Seasonal Variation in Zinc Concentration (ppm)

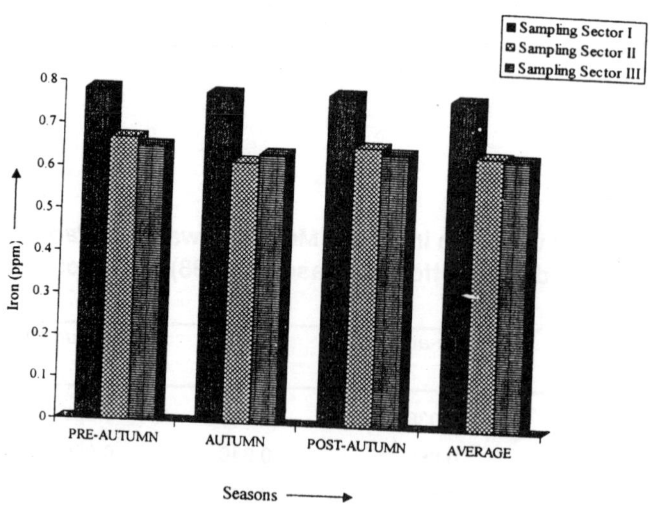

Fig. 5.18 : Seasonal Variation in Iron Concentration (ppm)

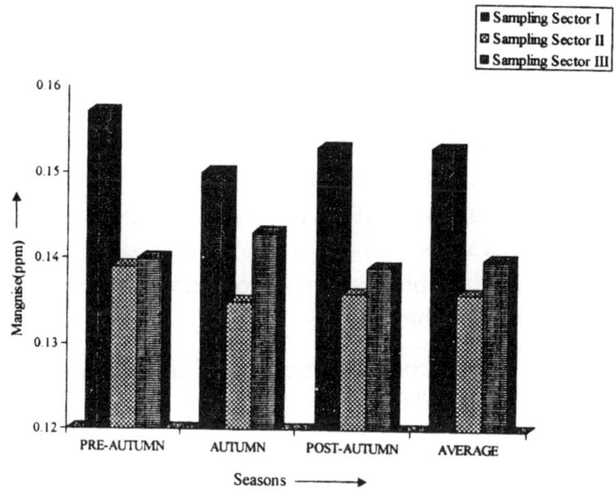

Fig. 5.19 : Seasonal Variation in Manganise Concentration (ppm)

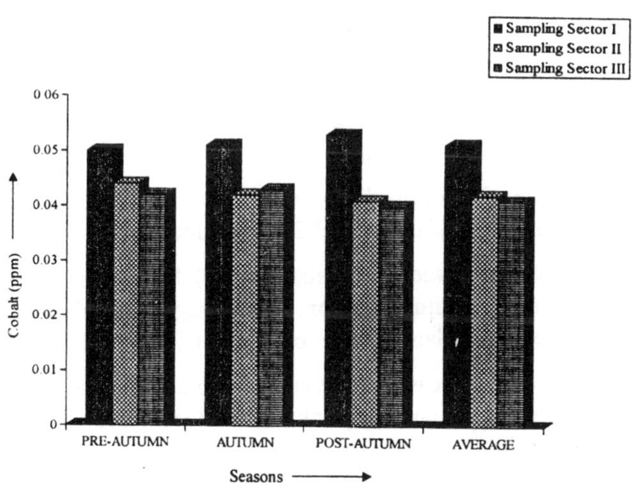

Fig. 5.20 : Seasonal Variation in Cobalt Concentration (ppm)

(D.1) Plankton

SAMPLING SECTOR - I

Quantitative analysis of plankton at sampling sector-1 revealed that the total planktonic concentration was highest (2196 units/1±149.10) is post-autumn season and lowest (1591 units/1±278.60) in pre-autumn season. The phytoplankton (1960 units/1±126.50) in post-autumn season and (1416units/1±258.00) in pre-autumn season. The zooplankton (236 units/l±22.62) in post-autumn season and (175 units/1±20.50) in pre-autumn season. The average of total plankton was (1985 units/1±729.72) contributed by both phyto-plankton and zooplankton (Tables 5.10 & Figs. 5.19) the seasonal percentage of plankton among chlorophyceae, Bacillariophyceae and Zooplankton were (60.4%, 21.7%, 7.0%, 10.5%), (Figs. 5 .21 to 5.22 and Tables 5.10).

Among phytoplanktonic concentration the Chlorophyceae was found higher followed by Bacillariophyceae and Cyanophyceae.

Maximum concentration of Chlorophyceae was (1376 units/1±145.60) in post-autumn season and (915 units/l ± 180.3) minimum in pre-autumn season. The average value of Chlorophyceae was (1170 units/l ±191.50).

The value of Bacillariophyceae was observed minimum (321 units/l ± 4.20) during pre-autumn season and maximum (482 units/l ±118.08) during post-autumn season. The average value at this sector was observed (315 units/l±127.73).

The minimum concentration of Cyanophyceae at this sector was (102 units/l ±19.09) in post-autumn season and maximum (180 units /1±36.06) in pre-autumn season. The average value was 129 units/1±35.68.

SAMPLING SECTOR-II

In sampling sector-II at Padali-Gujjar Pond, quantitative analysis of plankton revealed that. The total planktonic value was lowest (2044 units/1±28.99) during autumn season and highest (2138 units/1±37.47) in post-autumn season.

The maximum value of phytoplankton was observed (1907 units/l±23.33) in post-autumn season and minimum (1827 units/1±33.23) in autumn season.

The zooplankton was observed (231 units/1±14.14) during post-autumn season and (187 units/l ± 16.97) during pre-autumn season. The average of Total plankton was (2085 units/1±39.03) contributed phytoplankton and zooplankton (Table 5.11 & Fig. 5.23).

At this sampling sector the seasonal percentage of plankton among Chlorophyceae, Bacillariophyceae, Cyanophyceae, and Zooplankton were (58.7%, 25.7%, 8.7% and 10.1%) found respectively (Fig. 5.24).

The higher concentration of chlorophyceae was observed among phytoplankton followed by Bacillariophyceae and Cyanophyceae.

The lowest value of Chlorophyceae was observed (1079 units/1±110.30) in pre-autumn season and highest (1357 units/1±91.92) in post-autumn season. The average was recorded 1227 units ± 118.19.

The concentration of Bacillariophyceae was calculated minimum (438 units/1±21.21) during post-autumn season and maximum value was observed (527 unit/1± 41.71) during pre autumn season. The average value of Bacillariophyceae at this sector was 468 units/1± 4.27.

Highest value of Cyanophyceae was (291units/1±17.90) in pre-autumn season and lowest (112units/1± 46.66) in post-autumn season. The average value at this sector was observed 178 units/1±79.89.

SAMPLING SECTOR-III

At sampling sector-III of Padali -Gujjar Pond, quantitative analysis of plankton revealed that, the total planktonic concentration was maximum (2288units/1±214.25) during post-autumn season and minimum was recorded (1769 units/l ± 152.73) during pre-autumn season.

The highest value of phytoplankton was observed (2041 units ±192.33) in post-autumn season and lowest (1576 units/l ± 136.47) in pre-autumn season.

The zooplankton was recorded maximum (247 units/1 ± 21.92) during post-autumn season and minimum value was (193 units/1±16.26) during pre-autumn season. The average value of zooplankton at this sector was recorded 216 units/1 ± 22.64.

The average of total plankton was (1985 units/1±220.24) contributed by phytoplankton and zooplankton. (Table 5.12 Fig. 5.25). The seasonal percentage of plankton among, Chlorophyceae, Bacillariophyceae, and Cyanophyceae and Zoonplankton were (60.4%, 25.2%, 8.7%, 10.8%) (Fig.5.26 and Table 5.12).

The phytoplankton concentration at sampling sector-III of Padali-Gujjar Pond, followed by Chlorophyceae, Bacillariophuceae and Cyanophyceae.

Highest concentration of Chlorophyceae at this sector was observed (1479 units/1±192.33) during post-autumn season and lowest (1037 units/1±120.20) during pre-autumn season. The average value was calculated 1207 units/1±194.15.

Maximum value of Bacillariophyceae was recorded (432 units/l ±27.57) in autumn season and minimum was (324 units/1± 48.79) in pre-autumn season. The average was 393 units/1± 49.34.

The maximum value of Cyanophyceae at sampling sector-III was recorded (215units/1±33.23) during pre-autumn season and minimum (137units/1±21.96) during post-autumn season. The average value was recorded 168units/1± 34.49 at this sector (Fig.5.25).

(D. 2) Most Probable Number (MPN)

SAMPLING SECTOR-I

The lowest value of MPN at sampling sector-I of Padali-Gujjar Pond was recorded (151/100 ml ±30.12) in autumn season and highest value of MPN was recorded (230/100 ml ±25.7) in pre-autumn season. The average value of MPN stood at 193.6/100ml ± 32.5.

SAMPLING SECTOR-II

The maximum value of MPN at this sector was observed (202/100 ml $\pm$ 59.18) during pre-autumn section and minimum (60.6/100ml $\pm$ 40.80) during post-autumn season. The average value was 118.3/100 ml $\pm$ 60.58.

SAMPLING SECTOR-III

The value of MPN was recorded highest (96/100 ml $\pm$ 15.13) in pre-autumn season and lowest value of MPN at this sector was recorded (48.6/100ml $\pm$ 18.38) in post-autumn. The average was recorded 74.6/100 ml $\pm$ 41.94 (Table 5.13 and Fig.5.27).

(D. 3) Standard Plate Count (SPC)

SAMPLING SECTOR - I

The highest value of SPC at sampling sector-I was calculated (71×10^4 /ml $\pm$ 8.98) in pre-autumn season and lowest (41×10^4/ml $\pm$12.23) post autumn season. The average value was recorded 58.3×10^4/ml $\pm$ 12.68.

SAMPLING SECTOR-II

Maximum value of SPC was recorded (67×10^4/ml $\pm$ 9.89) during pre-autumn season and minimum (33×10^4/ml $\pm$14.12) in post-autumn season. The average value of SPC at the sector was calculated 53×10^4/ml $\pm$14.51.

SAMPLING SECTOR-III

The value of total count at sampling sector III was calculated maximum (52×10^4/ml $\pm$ 9.68) during pre-autumn season and minimum (25×10^4 /ml $\pm$ 9.40) during post-autumn season. The average was observed 38.3×10^4/ml $\pm$ 11. 10 (Table 5.14 and Fig. 5.28).

(D.4) Fishes and Economics

(a) Fishes

The Padali-Gujjar Pond is used for composite fish culture practice since very long time. During study period, 11 fish species are identified, among them seven fish species are culturable named *Catla-catla, Labeo-rohita, Cirrhinus-mrigala, Hypophthalmichthys-molitrix, Aristichthys-nobalis, Cyprinus-carpio, Ctenopharyngodon-idella* and five species are categorised as weed fishes namely *Channa-stratius, Glossogobius-giuris, Notopterus-notopterus* and *Chanda-nama*

The specimen's of the fishes were identified and observed by biometrical analysis, and fin formula of each culturable and weed fishes, showed in Tables 5.15, 5.16, 5.19 and Figs. 5.30 to 5.40 respectively.

On the basis of feeding habit, the culturable and weed fishes were divided into three main categories viz. Surface feeder species are *Catla-catla, Aristichthys-nobalis, Ctenopharyngodon-idella Channa-stratius, Glossogobius-giuris, Chanda-nama.*

Column feeder species are *Hypophthalmichthys-molitrix, Cyprinus-carpio* and bottom feeder species are *Labeo-rohita, Cirrhinus-mrigala, Notopterus notopterus.*

Table 5.10 : Qualitative fluctuation among plankton of *"Padali-Gujjar Pond"* Roorkee during different seasons (1998) at Sampling sector - I

S.No.	*Plankton*	*Pre- Autumn*	*Autumn*	*Post-Autumn*	*Average*
1.	Chlorophyceae	915±180.3	1221±36.00	1376±145.60	1170±191.50
2.	Bacillariophyceae	321±4.20	460±12.53	482±118.08	315±127.73
3.	Cyanophyceae	180±36.06	107±15.55	102±19.09	129±35.63
4.	Total phytoplankton	1416±258.00	1788±4.20	1960±126.50	1781±219.72
5.	Zooplankton	175±20.50	203±0.70	236±22.62	204±24.93
6.	Total plankton	1591±278.60	1991±4.24	2196±149.10	1985±729.72
Percentage					
1.	Chlorophyceae	57.5	61.3	62.6	60.4
2.	Bacillariophyceae	20.1	23.1	21.9	21.7
3.	Cyanophyceae	11.3	5.3	4.6	7.0
4.	Zooplankton	10.9	10.1	10.7	10.5

± = Standard Deviation

Table 5.11 : Qualitative fluctuation among plankton of *"Padali-Gujjar Pond"*, Roorkee during different seasons (1998) at Sampling sector - II

Sl.No.	*Plankton*	*Pre-Autumn*	*Autumn*	*Post-Autumn*	*Average*
1	Chlorophyceae	1079±110.30	1253±18.38	1357±91.92	1227±118.19
2	Bacillariophyceae	527±41.71	441±19.09	438±21.21	468±4.27
3	Cyanophyceae	291±17.90	133±31.81	112±46.60	178±79.89
4	Total phytoplankton	1889±10.60	1827±33.23	1907±23.33	1874±34.26
5	Zooplankton	187±16.97	217±4.24	231±14.14	211±18.36
6	Total plankton	2076±6.36	2044±28.99	2138±37.47	2085±39.03
	Percentage				
1	Chlorophyceae	51.5	61.3	63.4	58.7
2	Bacillariophyceae	23.3	21.5	20.4	25.7
3	Cyanophyceae	14.0	6.5	5.2	8.7
4	Zooplankton	9.0	10.6	10.8	10.1

± = Standard Deviation

Table 5.12 : Qualitative fluctuation among plankton of *"Padali-Gujjar Pond"*, Roorkee during different season (1998) at Sampling sector - III

Sl.No	*Plankton*	*Pre-Autumn*	*Autumn*	*Post-Autumn*	*Average*
1.	Chlorophyceae	1037±120.20	1106±71.41	1479±192.33	1207±194.15
2	Bacillariophyceae	324±48.79	432±27.57	425±22.62	393±49.34
3	Cyanophyceae	215±33.23	154±9.89	137±21.92	168±34.49
4	Total phytoplankton	1576±136.47	1692±54.44	2041±192.33	1769±119.96
5	Zooplankton	193±16.26	209±4.94	247±21.92	216±22.64
6	Total plankton	1769±152.73	1901±59.39	2288±214.25	1985±220.24
	Percentage				
1	Chlorophyceae	58.6	58.1	64.6	60.4
2	Bacillariophyceae	18.3	39.0	18.5	25.2
3	Cyanophyceae	12.1	8.1	5.9	8.7
4	Zooplankton	10.9	10.9	10.7	10.8

± = Standard Deviation

Table 5.13 : Seasonal variation in MPN of *"Padali-Gujjar Pond"* water, Roorkee during different seasons at Sampling sector - I, II, III (1998)

Sectors	*Parameters*	*Pre-autumn*	*Autumn*	*Post-autumn*	*Average*
Sampling Sector-I	(MPN/100ml.)	230±25.7	200± 4.52	151±30.12	193.6±3.56
Sampling Sector-II	(MPN/100ml.)	202±59.18	92.3±18.3	60.6±40.80	118.3±60.58
Sampling Secttor-III	(MPN/100ml.)	96±15.15	79.3±3.32	48.6±18.38	74.6±41.94

± = Standard Deviation

Table 5.14 : Seasonal variation in SPC of *"Padali-Gujjar Pond"* water, Roorkee during different seasons at Sampling sector-I, II, III (1998)

Sectors	*Parameters*	*Pre-autumn*	*Autumn*	*Post-autumn*	*Average*
Sampling Sector-I	(SPC/ml: 10^4)	71±8.98	63±3.32	41±12.23	58.3±12.68
Sampling Sector-II	(SPC/ml: 10^4)	67±9.89	59±4.24	33±14.12	53±14.51
Sampling Sector-III	(SPC/ml: 10^4)	52±9.68	38±0.21	25±9.40	38.3±11.10

± = Standard Deviation

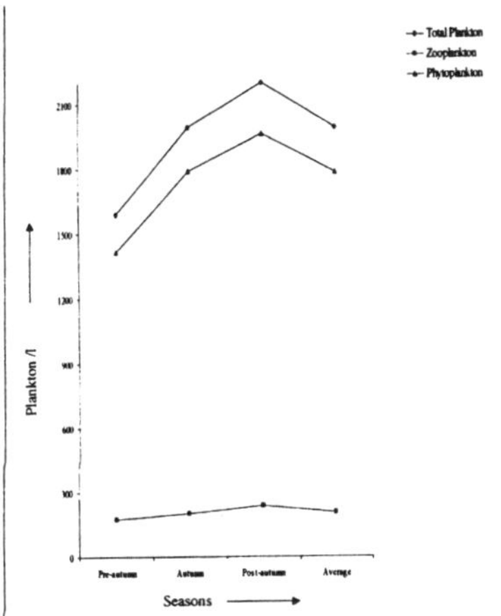

Fig-5.21: Seasonal Quantitative variation of Plankton at Sampling Sector I

Fig. 5.21 : Seasonal Quantitative Variation of Plankton at Sampling Sector-I

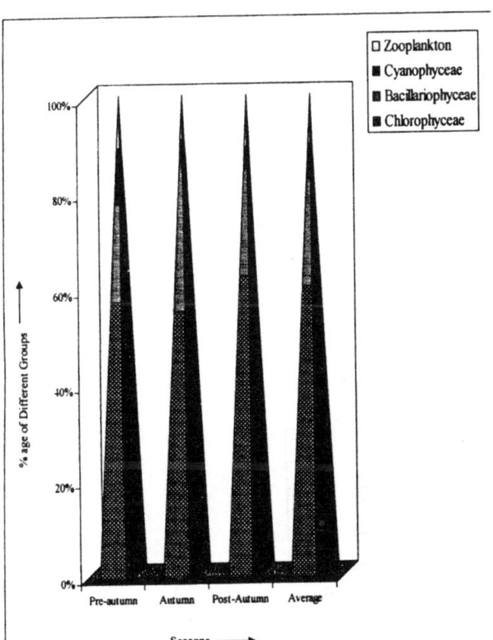

Fig. 5.22 : Quantitative Fluctuation in Plankton at Sampling Sector-I

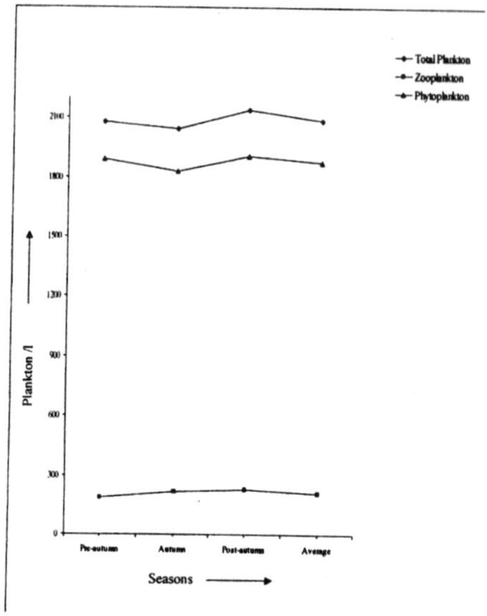

Fig. 5.23 : Seasonal Quantitative variation of Plankton at Sampling Sector-II

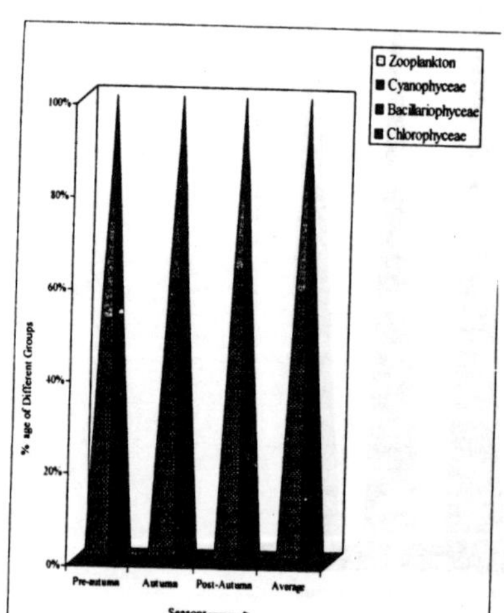

Fig. 5.24 : Quantitative Fluctuation in Plankton at Sampling Sector-II

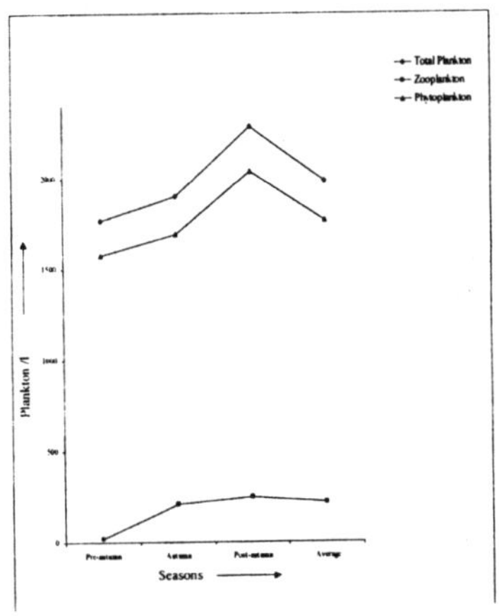

Fig. 5.25 : Seasonal Quantitative Variation of Plankton at Sampling Sector-III

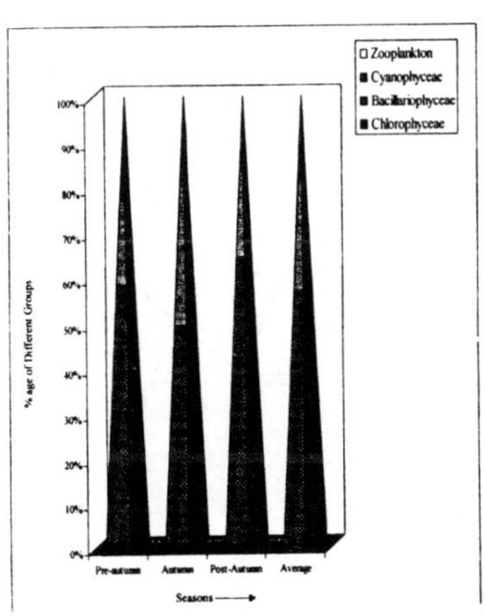

Fig. 5.26 : Quantitative Fluctuation in Plankton at Sampling Sector-III

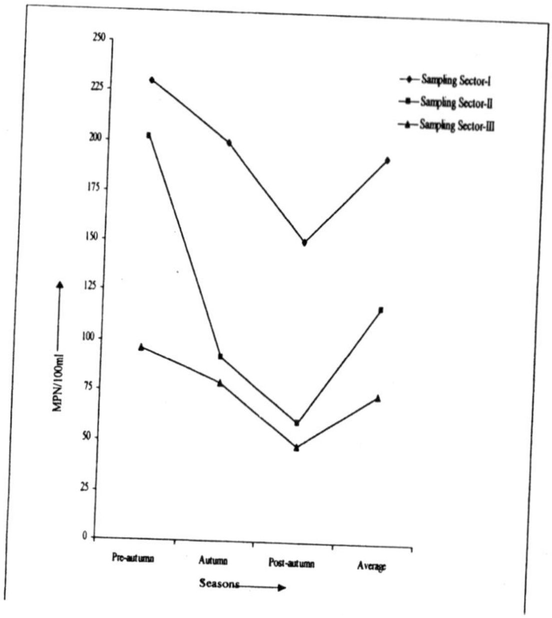

Fig. 5.27 : Seasonal Variation in MPN at Padali-Gujjar Pond

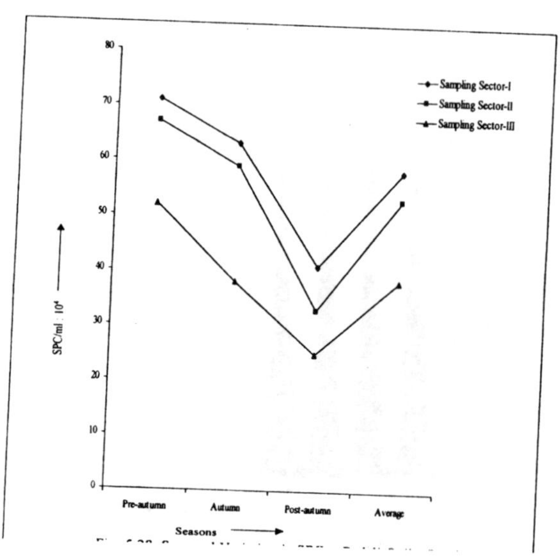

Fig. 5.28 : Seasonal Variation in SPC at Padali-Gujjar Pond

The pond have huge biomass of aquatic weeds and macrophytes in the shallow region of shore boundary.

The Grass carp *(Ctenopharyngodon-idella)* was found in maximum size and weight range was found (40-55 cm, 1.0 to 1.6 Kg.) respectively. The size weight range of other Culturable fishes are *Catla-catla* (35-40cm, 0.5-.830 Kg.), *Labeo-rohita* (35-45cm, 0.300-0.600 Kg.), *Cirrhinus mrigala* (30-35cm, 0.300-0.6 Kg.), *Hypophthalmichthys-molitrix* (50-60cm, 1.0-1.3 Kg.), *Aristichthys-nobalis* (45-55cm, 0.900-1.2 Kg.), *Cyprinus-carpio* (25-30cm, 0.400-0.800Kg.) Table 5.18.

In composite fish culture practice, the seed of seven culturable species such as : *Catla-catla, Labeo-rohita, Cirrhinus-mrigala, Hypopthalmichthys-molitrix, Aristichthys-nobalis, Cyprinus carpio, Ctenopharyngodon-idella* were stocked in the pond in ratio of 15:25:15:15:10:10:10 respectively. The seed of culturable species stocked from of April to August, 1997. The stocking seed size of all countable species ranges from 25-30mm, were supplied by *UP state fisheries departmem* (Table5.17).

At the time of harvesting of fishes, total productivity of culturable fish species were calculated with total size and weight ranges (fish productivity 1215Kg./ha/annum.).

The fish productivity and their percentage were *Catla-catla* (1048.0Kg., 17.96%), *Labeo-rohita* (1333.0Kg, 22.85%) *Cirrhinus-mrigala* (831.0Kg., 14.24%), *Hypophthalmichthys-molitrix* (600.0Kg., 10.28%), Aristichthys-nobalis (525.0Kg., 9.0%) *Cyprinus-carpio* (395.0Kg., 6.77%), *Ctenopharyngodon-idella* (650. 0Kg., 11.14%) showed in Table 5.18 and Fig. 5.29.

The fish productivity and the percentage of weed fishes found 450Kg., 7.71 % showed in Table 5.18.

(b) Fish Economics

Padali-Gujjar Pond is also a major source to earn money through composite fish culture in the present study the economy of composite fish culture during the year 1998 was studied in detail (Table 5.20).

The total expenditure on pond management was calculated Rs. 12,015 which contributed by maintenance of pond embankment/ outer boundries and in the removal of weeds and aquatic vegetation was Rs. 1000 and Rs. 1890 respectively. $CaCO_3$ and common salt were used to purify the pond water and the expenditure on them was Rs. 3000 and Rs. 6125 respectively.

The expenditure on seed stockment on seed viz. on seed stockment Ms Rs. 7,500 on the rate of 75/1000 seed.

The expenditure at the time of harvesting practices was Rs. 2370 for net operation, labour charges, Rs. 800 for rent of nets and Rs. 1200 on transport of fishes to local market.

The total expenditure in all the practices for composite fish culture was Rs. 23,885 and the total fish production was of Rs. 1,74,596.50 at the Padali-Gujjar Pond at the year 1998.

At last the net profit to owners of pond was Rs. 1,50,711.50 during 1998.

Table 5.15 : List of existing culturable fish species in *Padali-Gujjar Pond*

Sl. No.	Name (Generic-Species)	Common names	Family	Observed fin-formula
	CULTURABLE FISHES			
1	Catla-catla	Catla	Cyprinidae	D-16,P-7,V-8,A-7,C-21, Ll-37,Ltr-6/4
2	Labeo-rohita	Rohu	Cyprinidae	D-14,P-15,V-8,A-7,C-18,Ll-41,Ltr-7/7
3	Cirrhinus-mrigala	Mirgala	Cyprinidae	Barb (1Pair) D-14, P-1 1, V-8,A-6, C-23, Ll-41, Ltr-8/7
4	Hypophthalmichthys -molitrix	Silver carp	Cyprinidae	D-8, P-14, V-8, A-12, C-14, Ll-99, Ltr. 21/13
5	Aristichthys-nobalis	Big head	Cyprinidae	D-9, P-17, V-8, A-14, C-22, Ll-81, Ltr 17/21
6	Cyprinus-carpio	China carp	Cyprinidae	Barbados (Pair) D-10, P-16, V-10, A-7, C-18, Ll-31, Ltr-5/6
7	Ctenopharyngodon-idella	Grass cutter	Cyprinidae	D-9, P-16, V-9, A-9, C-23, Ll-35, Ltr-6/8

Table 5.16 : List of existing weed fishes species in *Padali-Gujjar Pond*

Sl. No.	Name (Generic-Species)	Common names	Family	Observed fin-formula
1	Channa-stratius	Dolly	Ophiocephalidae	D-24, P-16, V-5, A-20, C-11, Ll-37, Ltr-6/5
2	Glossogobius-giuris	Gulwa	Gobiidae	D-16, P-18, V-7, A-9, C-17, Ll-35, Ltr-4/3
3	Notopterus-notopterus	Dhuta	Notopteridae	D-7, P-15, V-7, A-95, C-23, Ll-222, Ltr-28/55
4	Chanda-nama	Chandla	Centropomidae	D-17, P 11, V-5, A-20, C-I6, Ll-scales deciduous

Table 5.17 : Stocking seed of culturable species in *Padali-Gujjar Pond* during 1997-98

S. No.	Culturable Species	Stocking month	Stocking ratio (%)	Stocking Seed Size
1	Catla catla	April	15	25-30mm.
2	Labeo-rohita	August	25	25-30mm.
3	Cirrhinus-mrigala	August	15	25-30mm.
4	Hypophthalmichthys-molitrix	August	15	25-30mm.
5	Aristichthys-nobalis	April	10	25-30mm.
6	Cyprinus-carpio	August	10	25-30mm.
7	Ctenopharyngodon-idella	April	10	25-30mm.

Table 5.18 : Harvesting of fishes during 1997-98 at *Padali Gujjar Pond,* Roorkee

Sl.No.	Culturable species	Size Range (CM)	Weight Range (Kg.)	Total Production (Kg.)	Percentage (%)
1	Catla-catla	35-40	0.5-0.830	1048.0	17.96
2	Labeo-rohita	35-45	0.8-1.0	1333.0	22.85
3	Cirrhinus-mrigala	30-35	0.3-0.6	831.0	14.24
4	Hypophthalmichthys-molitrix	50-60	1.0-1.3	600.0	10.28
5	Aristichthys-nobalis	45-55	0.9-1.2	525.0	9.00
6	Cyprinus-carpio	25-30	0.4-0.8	395.0	6.77
7	Ctenopharyngodon-idella	40-55	1.0-1.6	650.0	11.14
8	Weed fishes	5-15	0.050-0.202	450.0	7.71

Table 5.19 : Study of Biometric analysis of the fishes of *Padali-Gujjar Pond,* Roorkee during 1998

Sl. No.	Fishes species	Biometric parameters								
		Barbles	Eye-diameter (mm.)	Width of head through eye (mm.)	Width of body through operculum (mm.)	Width of body through dorsal fin (mm.)	Length of head (mm.)	Length of lateral line (mm.)	Total length of body (mm.)	Fin-formula
1.	Catla-catla	-	11	68	91	116	95	215	390	D-16, P-17, V-8, A-7, C-21, Ll-37, Ltr-6/4
2.	Labeo-rohita	-	10.5	45	70	90	75	175	310	D-14, P-15, V-8, A-7, C-18 Ll-41, Ltr-7/7
3.	Cirrihinus-mrigala	1 pair	10	35	52.5	465	55	190	315	Barb (1 pair) D-14, P-11, V-8, A-6, C-23, Ll-41, Ltr-8/7
4.	Hypophthalmichthys-molitrix	-	13	53	83	130	92	300	580	D-8, P-14, V-8, A-12, C-14, Ll-99, Ltr-21/13
5.	Aristichthys-nobalis	-	14	63	85	96	92	220	390	D-9, P-17, V-8, A-14, C-22 Ll-81, Ltr-17/21
6.	Cyprinus-carpio	1 pair	11	40	64	84	73	165	285	Barb(1 pair), D-10, P-16, V-10, A-7 C-18, Ll-31, Ltr-5/6 D-9, P-16, V-9 C-23, Ll-35, Ltr-6/8
7.	Ctenopharyngodon-idella	-	13.3	54	79.5	115	105	360	550	D-24, D-16,V-5, A-20, C-11, Ll-37, Ltr-6/5
8.	Channa-stratius	-	6	21	36	42	62	125	270	D-16, P-18, V-7, A-9, C-17, Ll-35, Ltr-4/3
9.	Glossogobius-Giuris	-	4	19	17	19	37	94	163	D-7, P-15, V-7, A-95, C-23, Ll-225, Ltr-28/55
10.	Notopterus-notopterus	-	8	23	38	54	40	145	195	D-17, P-11, V-5, A-20, C-16
11.	Chanda-nama	-	6	17	21	32	36	54	97	Ltr-scales deciduous

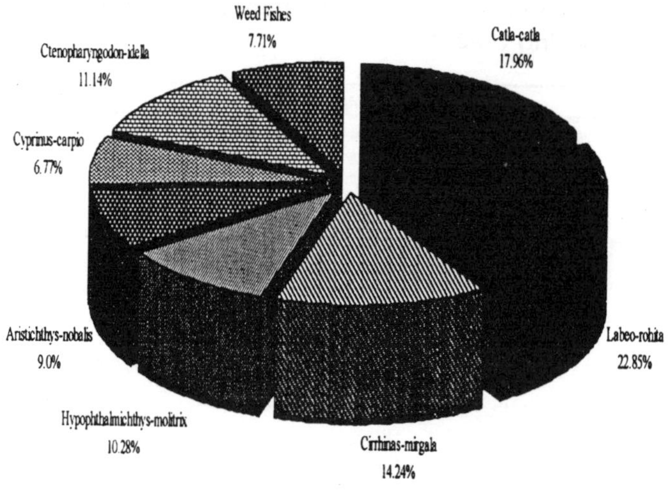

Fig. 5.29 : Total Production Percentage of Fishes at *Padali-Gujjar Pond* during April to December 1998

Table 5.20 : Economics of composite fish culture during 1998

Work	*Expenditure (Rs.)*
(A) Pond Management	
1. Maintenance of pond embankment/outer boundaries	= 1,000
2. Removal of weeds and aquatic vegetation[1] macrophytes	= 1,890
3. Calcium carbonates for water purification (150OKg. @ 2/Kg., two application/year)	= 3,000
4. Common salt (NaCl) (1750 Kg. @ 3.50/Kg., one time applied/year)	= 6,125
(B) Seed stockment	
1. No. of seed stocked (1,000,00) 20,000/ha/Annum@ 75/1000	= 7,500
2. Transport charges from Hatchery to fish pond (by govt. agency)	= Nil
(C) Supplementary feeds	
1. Supplementary feed not given	= Nil
(D) Harvesting practice	
1. Net (gillnet) operation Labour charges.	= 2,370
2. Rent of Nets	= 8,00
3. Transport charges of fishes from pond to Market	= 1,200
Total expenditure	= 23,885

Total fish production

Sl. No.	*Fish Species*	*Total Quantity (Kg.)*	**Rate/Kg*	*Total income (Rs.)*
1	Catla-catla	1048.00	30-55	34,060
2	Labeo-rohita,	1333.00	35-40	49,987.50
3	Cirrhinus-mrigala	831.00	30-32	25,761
4	Hypophthalmichthys-molitrix	600.00	22-25	14,100
5	Aristichthys-nobalis	525.00	20-25	11,812.50
6	Cyprinus-carpio	395.00	20-25	8,875.50
7	Ctenopharyngodon-idella	650.00	35-40	24,375
8	Weed fishes	450.00	10-15	5,625
	Total	5832.00		1,74,596.50

Total income (Rs.) **= 1,74,596.50**

Net profit = Total income - Total expenditure

Net profit = Rs. 1,74,596 - 23885 = 1,50,711.50

Net profit/Hectare = 31, 398/Ha/annum

*The rate of different fishes was given according to the cooperative society. (Matasya Jeev Palan Sahkari Samiti, Roorkee, U.P.)

Fig. 5.30 : *Catla catla* (Catla)

Fig. 5.31 : *Labeo rohita* (Rohu)

Fig. 5.32 : *Cirrhinus mrigala* (Mrigal)

Fig. 5.33 : *Hypophtalmichthys molitrix* (Silver)

Fig. 5.38 : *Glossogobius giuris* (Gulwa)

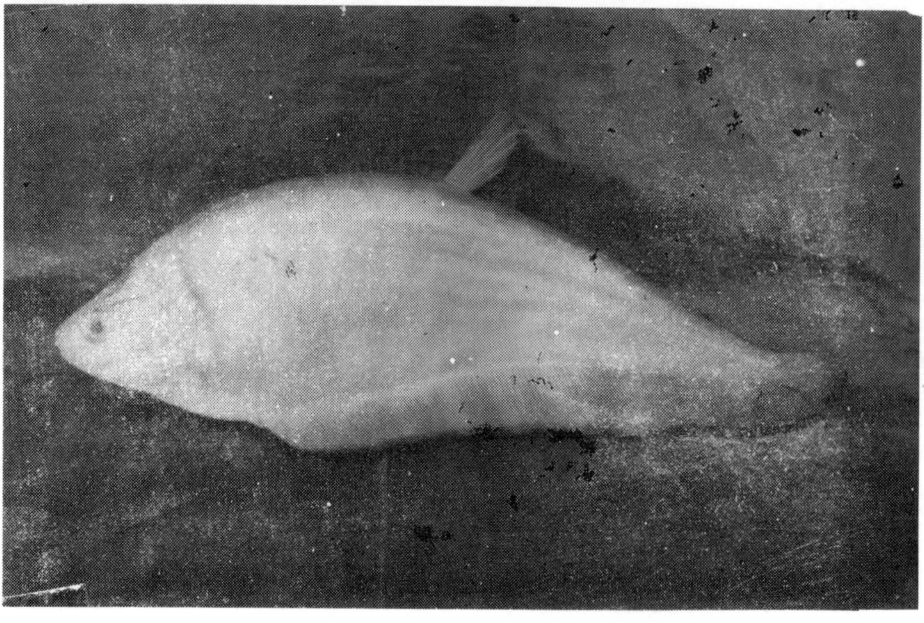

Fig. 5.39 : *Notopterus notopterus* (Dhuta)

Fig. 5.40 : *Chanda nama* (Chandla)

CHAPTER - 6

DISCUSSION

Padali-Gujjar-Pond (29°5' North Latitude and 77°53' East Longitude) is situated at an elevation of 268 meter above the sea level at Roorkee in district Haridwar. The surface area of the pond in 2,92,500 square meter (m^2) and catchment area 2,92,400 m^2.

The present study was done to study the broad limnological features of the pond with special reference to fisheries, during different seasons of year 1998 are tabulated in Tables 6.1 to 6.5 respectively.

The physico-chemcial and biological characteristics of water body play an important role in determination of the structure and composition of biotic community of any water body.

The characteristics are very much influenced by altitude, climate, vegetation, soil structure and cultural activities in the drainage are of the water body.

The temperature distribution in the water body is wholly dependent upon the thermal regime of water.

Productivity of a water body is influenced by the degree of variation in temperature. Temperature is often a limiting factor for the growth, reproduction and distribution of animals and plants, Michael (1986).

The water temperature of Padali-Gujjar Pond showed a decreasing trend during the study period. Similar observation was reported by Das and Srivastava (1956) from a tropical water body of North India.

The range of temperature was found to be conductive for the growth of Indian major carps by Jhingran (1982).

Residue is the solid matter present in water, water with high solids is harmful to organism, Michael (1986). Welch (1952) stated that dissolved solids vary qualitatively and quantitatively in different waters depending upon the seasons, locations and other factors.

Turbidity in the pond was fluctuated between 23.90 J.T.U. ± 32.97 to 108.33 J.T.U. ± 26.72 and showed a decreasing way. Similar findings have been observed by Ray and Banerjee (1978) and Sugunan (1980).

Conductivity in water represent the electrical conductance capacity of water, a decreasing way of conductivity observed during the study up to 188 micro mho ± 7.77 to 157 micro mho ± 4.12.

Table 6.1 : A Average value of Physical parameters of the water of *"Padali-Gujjar Pond"* during different seasons (1998)

Sl. No.	*Physical Parameters*	*Pre-Autumn*	*Autumn*	*Post-Autumn*	*Average*
1	Temperature (°C)	30.70±2.02	28.73±0.63	24.20±2.56	27.83±2.72
2	Total solids (ppm.)	314.73±24.71	275.28±3.42	250.11±21.22	280.13±26.78
3	Total dissolved solids (ppm.)	214.87±21.63	175.09±6.49	162.92±15.09	184.27±22.18
4	Total suspended solids(ppm.)	100.34±1.41	101.86±3.88	87.01±6.61	96.36±6.67
5	Turbidity (J.T.U.)	108.33±26.79	79.843±6.29	23.90±32.97	70.53±35.10
6	Conductivity (micro mho)	188±7.77	185±05.65	157±14.12	177.0±13.96

± = Standard Deviation

Table 6.2 : Average value of Chemical parameters of the water of *"Padali-Gujjar Pond"* during different seasons (1998)

SL. No.	*Chemical Parameters*	*Pre-Autumn*	*Autumn*	*Post-Autumn*	*Average*
1	PH	8.36±0.007	8.30±0.04	8.47±0.02	8.37±0.40
2	Dissolved oxygen (ppm.)	9.34±0.36	9.80±0.03	10.76±0.64	9.85±0.60
3	B.O.D (ppm.)	2.79±0.14	2.76±0.12	2.22±0.26	2.59±0.26
4	C.O.D (ppm.)	7.68±0.40	7.08±0.54	6.08±0.72	7.11±0.67
5	Free carbon-di-oxide (ppm.)	1.63±0.06	2.20±0.33	1.86±0.09	1.72±0.29
6	Alkalinity (ppm.)	73.22±0.24	71.51±0.96	73.31±0.31	72.87±0.84
7	Chlorides (ppm.)	59.54±2.24	55.08±0.90	54.46±1.34	56.36±2.26
8	Hardness (ppm.)	76.32±3.54	69.83±1.04	67.88±2.42	71.31±3.60
9	Calcium "Ca" (ppm.)	66.71±1.52	62.86±1.19	64.22±0.23	64.55±1.59
10	Magnesium "Mg" (ppm.)	9.54±1.14	6.96±0.67	7.28±0.45	7.92±1.14
11	Sodium "Na" (ppm.)	4.16±0.68	2.81±0.40	2.62±0.39	3.19±0.65
12	Potassium "K" (ppm.)	3.51±0.16	4.51±0.86	1.85±1.01	3.28±1.09

±= Standard Deviation

Table 6.3 : Average seasonal variation in heavy metals of water of *"Padali-Gujjar Pond"*, Roorkee during 1998

S.No.	*Heavy Metals (ppm.)*	*Pre-autumn*	*Autumn*	*Post-autumn*	*Average*
1	Zinc (Zn)	0.037	0.035	0.035	0.035
2	Iron(Fe)	0.701	0.679	0.699	0.693
3	Manganese(Mn)	0.145	0.142	0.142	0.109
4	Cobalt (Co)	0.045	0.045	0.044	0.044

Table 6.4 : Qualitative fluctuation among plankton of *"Padali-Gujjar Pond"*, Roorkee during 1998

S.No.	*Parameters*	*Pre-Autumn*	*Autumn*	*Post-Autumn*	*Average*
1	Chlorophycea	1010±135.05	1193±5.65	1404±143.5	1201+161.99
2	Bacillariophyceae	588±138.50	444±36.76	448±39.59	392+121.41
3	Cyanophyceae	228.66±49.96	131±19.09	117±28.99	158+49.67
4	Total phytoplankton	1627±127.98	1769±27.57	1969±113.84	1808 +141.66
5	Zooplankton	185±17.67	209±0.70	238±19.79	210+21.67
6	Total plankton	1812±145.66	1978±28.28	2207±133.64	2018+163.05
	Percentage				
1	Chlorophyceae	55.8	60.2	63.5	59.8
2	Bacillariophyceae	20.56	27.8	20.2	22.8
3	Cyanophyceae	12.40	6.6	5.2	8.1
4	Zooplankton	10.20	10.5	10.7	10.4

Table 6.5 : Average seasonal variation in MPN & SPC of water at *"Padali-Gujjar Pond"*, Roorkee during 1998

Parameters	*Pre-autumn*	*Autumn*	*Post-autumn*	*Average*
(MPN/100ml)	176±33.35	123.80±3.55	87.73±29.06	128.83±36.23
(SPC/ml:10^4)	63.33±9.52	53.33±2.45	33±11.92	49.86±12.61

± = Standard Deviation

Jhingran (1982) reported that pH is determined by the relationship between free CO_2 and carbonates. In the present investigation the pH value was observed low due to absence of carbonates and present of free CO_2. This is inconformity with the findings of Swarup and Singh (1978), Jhingran (1982).

Dissolved oxygen showed a increasing trend, in the present study in the pre-autumn season the DO was found to below due to higher metabolic rate of the organismic biomass at increased temperature. DO showed a inverse relation with temperature, as reported by Griffith (1955), Verma (1964).

BOD is the potential loss in oxygen due to decomposition of organic matter. BOD showed a negative relation with dissolved oxygen as reported by Verma *et al.* (1984).

The chemical oxygen demand in the Padali Gujjar Pond showed a decreasing trend during the study period. COD showed a positive relationship with BOD while showed a negative relationship with dissolved oxygen as observed by Chopra and Patrik (1994) in Ganga river at Rishikesh.

No conspicuous trend of fluctuation in free CO_2 was observe during investigation period in pond water as found by Rawat (1991) in Tumaria reservoir.

Tropical water bodies usually show a wide range of fluctuation in alkalinity, depending upon the location season, and planktonic population, Jhingran (1982). The cause of presence of alkalinity during this study was due to the presence of CO_2. Alkalinity above is supposed to be nutrient and such waters are often the best for fish culture as Barrett (1953).

Chlorides may occur in fresh waters as a result of dissolution of salts deposits in the soil, Michael (1986). The increase in chloride content was due to excess of suspended salts in water. The concentration of chlorides showed decreasing manner through out the study. Similar type of trend as observed by Singh (1960), and Munawar (1970).

Hardness indicated the concentration of Calcium and Magnesium ions. Padali Gujjar Pond with average hardness of 71.30 ppm±3.60 may be characterised as hard water body. Hardness showed a decreasing trend in the study and its value is satisfactory for fish growth.

The concentration of calcium in the Padali Gujjar Pond found to be maximum 66.71 ppm ± 1.52 in pre-autumn season. The pond may be regarded as rich in calcium content as marked by Dwivedi (1995) in two ponds at Patna.

Concentration of all the heavy metals detected in Padali-Gujjar Pond were quite favourable to the biota and not infavourable in the water quality as suggested by Mathur (1982).

Algae present in pond water because of photosynthetic activity, release oxygen in the water and this oxygen is available to increase aerobic decomposition of the organic wastage of bacteria.

The present study showed that the maximum number of total plankton was 2207 units/l ± 133.64 , phytoplankton was 1969 units[l] ± 113.84 and zooplankton was 238 units/l ± 19.79 in post-autumn season. The minimum concentration of total plankton was recorded to be 1812 units/l ± 145.66, phytoplankton was 1627 units/l ± 127.98 and zooplankton was 185 units/l ± 17.67 in pre-autumn seasons.

The results indicate that the number of plankton were affected due to temperature and turbidity of water.

The same type of results were also reported in different ponds by Hutchinson (1967), Das and Upadhyay (1979) in lakes of Kashmir and Nanital. The maximum percentage of different groups were noted as Chlorophyceae (63.5%) in post-autumn season, Bacillariophyceae (27.8%) autumn and Cyanophyceae (12.4%) in pre-autumn seasons.

In this study the Chlorophyceae was a dominating group. The number of plankton were also affected by the turbidity. The increased turbidity reduce the plankton production as also reported by Das and Pahtani (1978).

The standard plate count of pond water was recorded 49.86×10^4 /ml ± 12.61and MPN were 128.83/100ml ± 36.23. It was highest in pre-autumn season 63.33×10^4/ml ± 9.52 and 176/100ml ± 33.35 in SPC and MPN respectively due to high percentage of nutrients which increase the growth and multiplication of bacteria.

In present study of fish culture practice in Padali Gujjar pond revealed some specific information regarding the nature of fish culture, growth and fish production. The nature of fish culture were adopted by the farmer as Composite fish culture method with stocking seven culturable major carps species such as *Catla-catla, Labeo-rohita, Cirrhinus-mrigala, Hypophthalmichthys-molitrix, Aristichthys-nobalis, Cyprinus-carpio, Ctenopharyngodon-idella.* In perrenial pond the Indian major carps usually tolerate the wide range of temperature (18.3-37.8 °C). The grass carp *(Ctenopharyngodon-idella)* is known to stand in temperature up to 40.0°C Jhingran (1982). The pH range (8.36-8.47) of pond also exhibit the positive correlation with fish culture.

The waters having a pH range (7.5-9.0) provide the best medium for fish culture Swingle (1967a).

The total catch composition of culturable fishes have showed a correlation in between growth and feeding habit. The surface feeder, column feeder, and bottom feeder fishes had contributed 41.50%, 18.48% and 40.20% respectively. On the basis of total biomass of phytoplankton and zooplankton in pond, Das and Moitra (1955).

The grass carp *(Ctenopharyngodon-idella)* has contributed 11.14% of total fish production. In ponds four zones of vegetation are distinguished viz. Bottom zone, submerged rooted, unattached submerged, free floating and surface vegetation. The large biomass of aquatic vegetation occured in the pond e.g. *Vallisneria spiralsis, Hydrilla verticilata, Ofella alismoides, Najas, Potamogeton, Wolfia, lamena, Azolla, Pistia, Chlorella,-vulgarius* etc. Vishwas and Colder (1955).

The economics of composite fish culture of the pond has contributed fish productivity 1215 kg/ha/year, and the net profit was calculated Rs.31,398 per hectare/annum.

Finally all above observations and data indicated that a huge amount of proteins are achieved through conversion of biological waste material by composite fish culture.

CHAPTER - 7

SUMMARY

The book embodies a findings of study conducted on "Pond Fish Ecology & Economics Roorkee (Hardwar)" during three different seasons of year 1998 and consist of seven chapters. The text is supported by 26 tables and 46 figures including maps, graphs and photographs.

For this study the water samples were collected from three different sampling sectors. (Sampling sector–I, II and III. Water samples were analysed by the standard methods. The pond bed in all three sectors had clay in access. The results of water were made in Pre-autumn, Autumn, and Post-autumn seasons.

A summary of the findings is given below :

(1) The water temperature of the Padali Gujjar pond at Roorkee fluctuated between 24.20°C ± 2.56 to 30.70°C±2.02.

(2) The quantity of excess residue may affect the quality of water and also makes its potability of inferior nature. The value of total solids recorded between 250.11 ppm ±21.22 to 314.73 ppm ±24.71.

The value of total dissolved solids was found between 162.92 mg.1±15.09 to 214.87 ppm±21.63. The amount of total suspended solids ranged between 87.01 ppm ± 6.61 to 101.80 ppm ± 3.88.

(3) The maximum turbidity in pond water was observed between 23.90 J.T.U. ± 32.97 to 108.33 J.T.U. ± 26.70.

(4) The conductivity of the water fluctuated between 157 micro mho ± 14.12 to 188 micro mho±7.77.

(5) The pH of pond water found slightly alkaline during the study period and ranged between 8.30±0.04 to 8.47±0.02.

(6) The dissolved oxygen fluctuated between 9.34 ppm ± 0.36 to 10.76 mg./l ± 0.64.

(7) The BOD was maximum (2.79 ppm ± 0.14) and minimum (2.22 ppm ± 0.26) in pre-autumn and autumn seasons respectively.

(8) The chemical oxygen demand calculated minimum (6.88 ppm ± 0.72) in post-autumn season and (7.68 ppm ± 0.40) in pre-autumn season.

(9) The value of Free CO_2 fluctuated between 1.63 ppm ± 0.06 to 2.20 ppm ± 0.33.

(10) Alkalinity in the Padali Gujjar Pond water recorded between 71.51 ppm ± 0.962 to 73.22 ppm ± 0.24.

(11) The maximum concentration of chlorides was (59.54 ppm ± 2.24) in pre-autumn season and minimum (54.46 ppm ± 1.34) in post-autumn seasons.

(12) The excessive hardness of water makes the water of inferior quality. The hardness of pond water fluctuated between 62.85 ppm ± 1.19 to 66.71 ppm ± 1.52.

(13) The concentration of calcium in pond water recorded between 62.85 ppm. ± 1. 19 to 66.71 ppm ± 1.52.

(14) The magnesium concentration fluctuated between 6.96 ppm ± 0.67 to 9.54 ppm ± 1.14.

(15) Sodium and potassium both showed a decreasing trend during study period. Sodium concentration fluctuated between 2.62ppm ± 0.39 to 4.16ppm ± 0.69 and potassium concentration 1.85ppm ± 1.01 to 3.151ppm ± 0.16.

(16) The concentration of all heavy metals in pond water showed a decreasing trend. The concentration of Zinc (Zn) recorded between 0.035 ppm to 0.037ppm.

The iron concentration recorded between 0.679 ppm to 0.701 ppm. Manganese concentration found between 0.142 ppm to 0.145 ppm and the value of cobalt observed between 0.044 ppm to 0.045 ppm.

(18) The plankton of Padali Gujjar Pond showed a positive relation with dissolved oxygen, while showed a negative relationship with BOD, COD and temperature.

(19) The total coliform organism (MPN) of pond water at Roorkee ranged between 87.73/100 ml ± 29.06 to 176/100ml ± 33.55.

(20) The total count of bacteria (SPC) of the water ranged between 33×104/ml ± 11.92 to 63.33 × 104/ml ± 9.52.

(21) The fish fauna of the Padali-Gujjar Pond at Roorkee was surveyed in a detail. As a result of this survey 11 species belonging to 11 genera, 5 families.The biometric analysis of the survey fishes were described in text.

(22)The economy of composite fish culture practices in the Padali-Gujjar Pond was studied and final profit was found Rs. 1,50,711.50 during April to December, 1998.

The study conclude that the fish production at Padali Gujjar Pond during 1998 was found very low due to non scientific fish culture. It is suggested, to increase the fish production the proper scientific methods should be applied in pond.

BIBLIOGRAPHY

Adholia N. (1991). Primary productivity of Mansarovar Reservoir, Bhopal. *Him. J. Env. Zool.,* 5, 1991:79-8.

Ahmad N. (1952). Stocking of fish in East Pakistan. *Agric. Pakist.,* 3(4): 1-3.

Akhtar, S. And Das, S.M. (1970): A report on fresh water plankton from Dal lake Kashmir. *Kashmir. Sci.,* 7(1): 133-137.

Alikunhi, K.H. (1956a). Fish culture technique in India. *Proqr. Fish Developm. India.,* 63-73.

APHA, AWWA and WPCF (1980). *Standard methods for the Examination of water and waste water.* American Public Health Association. Washington 14th Edition. N. York pp. 193.

Bagde, U.S. and Verma A.K. (1985). Physico-chemical Characteristics of water of J.N.U. Lake at New Delhi. *Indian. J. Ecol.,* 12(l): 151-156.

Balone, EX and Coche, A.G. (1974). Lake Kariba, a man made tropical ecosystem in central Africa. *Magrbial,* 24:767.

Barrett, P.H. (1953). Relationship between alkalinity and adsorption and regeneration of added phosphorus in fertilized trout lakes. *Trans. Amer. Fish. Soc.,* 82: 78-90.

Bhuyan, B. R. (1970). Physico-chemical quality of water of some ancient tank in Sibsagar, Assam. *Ind. J. Env. HIM.,* 12:122-134.

Brige, E.A. and Juday, C. (1911). The inland lake of Wisconsis. The dissolved gases and their biological significance. *Bull. Wis. Geat. Nat Hist. Surv.,* 22: 259.

Byars, J.A. (1960). A fresh water pond in Newzealand. *Aust. J. Mar. Fresh. Res.,* 2(2): 222-240.

Chacko, P.I. and Krishnamurthy, B. (1954). On the plankton of three fresh water fish ponds in Madras City. *India. Symp. Mar. and Fresh water plankton India-pacific covn, Bangkok.* (UNESCO),103-107.

Chopra, A. K. and Patrik, Nirmal J. (1994). Effect of domestic sewage on self purification of Ganga water at Rishikesh. 1. Physico-chemical parameters. *Ad. Bids.* Vol. 13(l):75-82.

Choudhary, H. (1960b). Contributions to the technique of pond fish culture in India. *D. Phil. Thesis, Univ. of Calcutta,* Calcutta.

Cook, J. (1977). Environmental pollution by Heavy metals. *Internet, Environmental studies* 9:253-256.

Daborn, G. R. (1976). Physical and chemical factors of vernal temporary pond in western Canada. *Hydrobiologia* SI (1) : 33-39.

Das S.M. and Akthar, S. (1970). A report on fresh water plankton from Dal lake, Kashmir. *Kashmir Sci.,* 7(l)., 133-137.

Das S.M. and Moitra S.K. (1955). Studies on the food of some common fishes of Uttar Pradesh, India. 1. The surface feeders, the mid feeders and the bottom feeders. *Proc. nat. Acad. Sci India* 25B (1 and 2) :1-6.

Das, S.M. and Pathani, S. S. (1978). A study of the effect of lake ecology on productivity of Mahaseer (Tor-tor and Tor putitora) in Kumaon lakes, India. *Matsya,* 4:25-31.

Das S.M. and Shrivastava, V.K. (1956). Some new observations on plankton from fresh water ponds and tanks. *Lucknow. Sci. and Cult.,* 21(8) : 466-467.

Davis, N.C.(1932). The effects of heat and cold upon Aedes stegomyics aegypti. *Amer. J. Hyg.,* 16 :177-191.

Day, F.(1879). *The fishes of India,* Burma and Ceylon, being a natural Dawson, London, pp. 788.

Dutta, N.C., Mandal, N. and Bandhyopadhyay, B.K. (1987). Seasonal abundance of Rotifers in perennial fresh water pond in Calcutta. *J. Env. Biol.* 8 (1): 63-72.

Dwivedi,H.N. et al. (1995). limmological studies in relation to fish production of two ponds at Patna. *Itima. J. of Enviornment and Zoology,* 9(l):7-1 0.

Faruqi, A.J. and Sahai, R. (1943). Method of Catching fish in United provinces and scope of establishment of Indand Fisheries in these provinces. *Proc. Acad. Sci. India.* 133B(3): 198-214.

Forel, F.A. (1892). Le Leman. *Monographic Limnogique, 3 vols, Lavsane.*

Gaur, K, Komal. (1992). Physico-chemical and biological characteristics of Keetham lake. *J. Nat.,* 4(1) : 35-39.

Golterman, H.L. (1969). Methods for chemical Analysis of Fresh water. *IBP Handbook No. 8 Oxford: Black well Scientific.*

Griffith, R.S. (1955). Analysis of plankton yields in relation to certain physical and chemical factors of lake Michigen. *Ecology,* 36(4): 343-552.

Gunther, S.D. *rer nat* Habli (1959). *Fresh water fishes of the world. Vol. 1 & 2.* Translated by Tucker, D.W. (1973). T.F.H. Publication Inc. Ltd, London.

Hart, G.E. and Fuller S.L.H. (1974). Pollution Ecology of Fresh water invertebrates. *New York: Academic press.,* 230 pp.

Hickling, C. F. (1962). Fish Culture. 295pp. *London: Faber and Faber.*

Hickling, C. F. (1971). Fish Culture, 2nd ed : 317 p. *London. Faber and Faber.*

Hutchinson, G.E. (1957, 1967, 1976). *A Treatise of Limnology Vol I,* Geography, Physics and Chemistry of Lakes 1015 pp. Vol II. introduction to lakes Biology and Limnoplankton 1150 pp. Vol III, Aquatic communities and Energy Relationships. 1040 pp. New York wiley.

Hynes, H.B.N. (1970). The Biology of Polluted waters. Liver poll. *The university press,* 202 pp.

Ibrahim, K. H. (1957). Bionomics of forage fishes: observation on the fecundity of three common species of minor barbles. *J. Bombay Nat. Hist. Soc.54* (4): 826-34.

Jayaram, K.C.(1987). The fresh water fishes of India, Pakistan, Bangladesh, Burma and Srilanka. *Zoological Survey of India, Calcutta,* p. 480.

Jhingran, V.G. (1982). Fish and Fisheries of India. *Hindustan Publishing Corporation (India),* 1-954.

Jindal, R. and Ghezta, R. (1998). Pollution status of water bodies of Chandigarh. *Abs. Nat. Con. Sus. Eco Env.,* 68.

Joshi, B.D., Bisht, R.C.S. Joshi, Namita and Singh, Rajendra (1991). A study of planktonic and benthic components of three selected tributaries of river Ganga between Devprayag and Rishikesh. *Him. J. Env. Zool Vol. 10,* pp. 23-26.

Khan, H. (1930). Methods of fishing in the Punjab. *J. Bombay. Nat. Hist. Soc.,* 24.

Khanna, D.R. (1993). Ecology and pollution of the river Ganga *Ashish Publishing house,* Delhi, 1 241.

Khanna, D.R., Malik, D.S. and Badola, S.P. (1997). Population of green algae in relation to physico-chemical factors of the river Ganga at Lal-Ji Wala Hardwar. *Uttar Pradesh J. Zool., 17(3): 237-240.*

Kumar, Vijay K. paul, Ravindra and Grish Kadade Varu (1992). Effect of Methane parathion. On primary productivity of fresh water ponds. *J.Karnataka Sci.* 4(1): (95-97).

Lind, O.T. (1974). Hand Book of common methods. *limnology. St. Louis C. V. Mosby & Co.,* 154.pp.

Maar, A. (1966). Mortimer, M.A.E. and Linqen, 1. Vander (1966) Fish Culture in central East Africa, 160 p. Rome, FAO.

Macan, T.T. (1974). Fresh Water Ecology, 2nd ed. New York: Wiley, 224,pp.

Mahajan, C.L. (1961). Fish fauna of Muzaffarnagar district Uttar Pradesh. *J. Bombay Nat. Hist. Soc.,* 62(3) : 440-454.

Mathur, R.P. (1982). Water and waste water testing. *Nem Chand and Bros. Publishers,* Roorkee, 1-54.

Michael, P. (1986). Ecological methods for field and laboratory Investigations. *Tata McGraw Hill Publishing Company Limited, 1214 Asaf Ali Road, New Delhi.*

Munnawar, M. (1970). Made study on some fresh water pond at Hyderabad, *India. Biot Hydrobiologia.,* 35(1) : 127-162.

Nakamura, K. (1954). Fish production in seven form ponds in chicda Plain, Nagano Prefecture with reference to natural limnological environment and artificial treatment. *Bull. Fresh W. Fish. Res. Lab.* Tokyo, 3(l):79.

Natarajan, A.V. (1974). Ecology and fish culture in the reserviors of India with special reference to Rihand Nagar - Juno Sagar, Bhavani Sagar, Konar and Tilaiya. Summer institute or Inland Aqu aculture. *Central Inland fisheries research Institute, Barrackpore 27 June to 31st July 1974.*

Needham, J. G. And Needham, P. R. (1972). A guide to the study of fresh water Biology. *Holden-Day 1. N. C San Francisco, Colit.* 94 (3) : 1-108.

Padmavati, P. and Durga Prasad, M.K. (1997). Impact of unwanted fish on the water quality, plankton and fish production in a carp culture pond in the environ of lake Kalleru. J. Environ. Biol., 18(1): 79-93.

Paka, Swarnlathat Paka and Rao, A. Narsing (1997). Inter relationship of physico-chemical factors of pond. *J. Environ. Biol,* 18(l): 67-72.

Palmer, C. M. (1980). *"Algae and water pollution".*

Pant, M.C. (1970). Fish fauna of the Kumaon hills. *Rec. Zoo. Surv. India,* 64 (1-4): 85-96.

Peenak,R.W. (1962). Quantitative zooplankton sampling in Littoral vegetation areas. *Limnol. Oceanogr,* 7: 487-489.

Prasad, D.Y, (1990). Primary productivity and energy flow in Upper Lake, Bhopal *Indian. J. Env. Helth,* 32, 2: 132-139.

Prescott, G.W. (1972). The Algae A Review. *Riverside studies in Biology.41,* 280 pp.

Pruthi, H.S. (1933). Studies an the bionomics of fresh waters of India. I. seasonal changes in the physical and chemical conditions of the water of the tank in the Indian museum compound. *Int. Rev. Hydrobiol.,* 28: 46-67.

Rao, N.G. and Durve, V.S. (1989). Cultural eutrphication of the lake Ranga sagar, Udaipur, Rajasthan. India. *J. Environ. Biol.* 10(2), 127-134.

Rawat, H.S. (1991). Studies on the limnology and fisheries of Tumaria reservior at Kashipur, Nanital. *D. Phil Thesis.*

Ray, P. and Banerjee, R.K. (1978). Hydrological studies of Tungabhadra reservior. *Proc. On the ecology and fisheries of fresh water reserviors C.I.F.R.I. Barrack pove, Nov.* 27-29 : 129-154.

Reid, K.G. (1961). Ecology of Inland waters and Estuaries. *New York: Reinhold,* 280pp.

Ruttner, F. (1968). Fundamentals of Limnology. *Toronto: Univ of Toronto Press.* 332pp.

Sahai, R. singh, R.S. and Saxena, P.K. (1986). Effect of fertilizer factory effluent on the distributional pattern of Zooplankton in Chilka lake. *Ind. J. Bot.* 9(1) : 80-84.

Sehgal, J.K. (1973). Fish and fisheries of Aligarh with notes on the fish catche. *J. Inland Fish Soc. India. Vol.* V :23-28.

Saxena, M.M. (1994). Environmental Analysis water, Soil and Air. *Agro Botanical Publishers (India)* : pp 15-120.

Scheimer, F. (1983). Limnology of Parakrama samudra, Sri Lanka a case study of an ancient man made lake in the tropics. *Development in Hydrobiologoy,* 12:236.

Sharma, R.C. Juyal C.P. Gusain, O.P. and Rawat, M.S. (1993). First report on the Limnology: Abiotic profile of a Garhwal Himalayan Lake, Deoriatal. *Advances in limnology, Ed. H.R. Singh. Narendra publishing House, Delhi*, 87-92.

Singh, Room and Mahajan, I. (1987). Phytoplankton and water chemistry of Renuka lakes, Himachal Pradesh. *Ind. J. Ecol., 14(2): 273-277.*

Singh, S.B. and Sahai, K. (1986). The primary production of phytoplankton of a tropicial shallow water body of Gorakhpur. *Tropical Ecol.,* 27:33. 39.

Singh, S. P. (1974). Fishes of Moradabad (U. P.). *Ind. J. Zool.,* 15(2) : 75-78.

Singh, K.P. (1960). The chlorophyceae of Kumaon Hills, U.P. India. *J. Proc.Ind. Acad. Sci.* 161-196.

Sladecek, J. (1963). A guide tc Limno-Saprobial crganisms. *Sci. Inst. Tech. Progr. Technol.* 7: 543-562.

Sugunan, V.K. (1980). Seasonal fluctuations of Plankton of Nagar Junasagar reservior A. P (India). *J. Inland Fish Sac. India,* Vol, 12(l) : 79-91.

Swarup, K. and Singh, S.R. (1979). Limnological studies of Suraha lake (Ballia). *J. Inland Fish Soc. India.* 2(l): 22-23.

Swingle, H.S. (1967a). biological means of increasing productivity in ponds. *F.A.O. Fish-Rep* (44): 243-57.

Talwar, P.K. and Jhingran, (1991). Inland fishes of India and Adjacent countries. Vol. I & II Oxford IBH Publishing Co. Pvt. Ltd.

Tombi, Singh, H. (1992). Ecobiology of Hill stream fishes of Manipur with special reference to culturable possibilities. *Final Technical Report. DSTE, Govt. of Manipur,* pp. 1-50.

Tripathi, S.D. (1990). Fresh water aquaculture in India. In Joseph M.M. (ed). Aquaculture in Asia. *Asian Fisheries Society, Indian Branch, Mangalore,* 1-396.

Trivedi, R.K. and Goel, P.K. (1984). Chemical and Biological Methods for water pollution studies. *Karad Environmental Publication*, 1-25.

Verma, J. P. and Mohanty, R. C. (1995). Phytoplankton and its correlation with certain physico-chemical parameters of Damnukundpur pond. *Environmedia Pollution Research,* 14 (2): 233-242.

Verma, S.R. and Shukla, G.R. (1970): physico-chemical conditions of Kamla Nehru Tank Muzaffernagar (U.P.) in relation to biological productivity. *Ind. J. Env. Hlth.* 12: 10-128.

Verma, M. (1964). Hydrobiological survey of the Dalsagar tank, saloni, M.P. India; with special reference to the effect of Microcystis aeruquinosa (Kutza), bloom on fish production and suggestion for the improvement of the tank fishery. *Icthyologica.* 3:37-46.

Verma, S.P., Sharma, P., Tyagi, A., Rani, S., Gupta, A.K. and Dalela, R.C. (1984). Pollution and saprobic status of eastern Kalinadi. *Limnological (Berlin).* 15(1) : 69-113.

Vishwas, and Colder (1955). Study on the limnology of the aquatic vegetation.

Ward, H.B. and Whipple, G.C. (1966). Fresh water Biology. John Willey and Sons, Inc. New York: 1-1203.

Welch, P.S. (1948) Limnological Methods. New York, McGraw-Hill 381 pp.

Yang, S.L. (1982). Fish Culture and reservoir Management. in the Republic of Singapore. In Seminar on production and esxploitation of open waters. *Regional centre for Tropical biology, Bogor Indonesia,* 1-18.

Yashov, A. (1959b). Studies on the productivity of fish ponds. 1. Carrying capacity S. *Proc. tech. pap. gen Fish. Coun, Mediterr S.* 407-19.

Yousuf, A.R. and Quadri, M.Y. (1978). Seasonal Variation in the physico chemical factors of Subtropical lake in Kashmir. *Ind. J, firsh Soci. India.* 10: 86-96.

Zafar, A.R. (1964). On the ecology of algae in certain fish ponds of Hyderabad, India. III. The periodicity. Hydrobiologia, 30(l): 96-112.